THE NATURE OF SOCIAL SCIENCE

The
Nature of
Social Science

GEORGE C. HOMANS

AN ORIGINAL HARBINGER BOOK
HARCOURT, BRACE & WORLD, INC.
NEW YORK

PREFACE

The three chapters of this book were delivered, in earlier versions, as Walker-Ames Lectures at the University of Washington in the summer of 1965. I wish to thank the President and Board of Regents of the University for the opportunity to give these lectures; also the members of the Department of Sociology, especially Professor Robert E. L. Faris, and other members of the university staff for making my summer in Seattle a stimulating one, intellectually and otherwise.

George C. Homans

CONTENTS

CONTENTS

I had looked for a society reduced to its simplest ex-pression. That of the Nambikwara was so far reduced that I found only men there.

—CLAUDE LÉVI-STRAUSS
Tristes Tropiques

1
DISCOVERY
AND EXPLANATION

For our purposes—my readers' and mine—the social sciences include psychology, anthropology, sociology, economics, political science, history, and probably linguistics. These sciences are in fact a single science. They share the same subject matter—the behavior of men. And they employ, without always admitting it, the same body of general explanatory principles. This last truth is so obvious that it is still highly controversial.

The question most often asked of these sciences is the

one question I shall not ask: whether they are sciences at all. The very effort to answer it by demonstrating in action that they are truly no-nonsense, brass-instrument, experimental scientists has damaged some sociologists: they have gotten diverted from the matter of science to the manner. Yet in all these fields even scholars less preoccupied with their status would have no trouble agreeing that they were scientists—except perhaps the historians. The historians get the best of both worlds: they become humanists when judged by the scientists and scientists when judged by the humanists. Not all the physical scientists, of course, would agree that the social sciences were sciences. They would argue that social science was not exact and could not make many specific predictions. By these standards Darwin's theory of evolution would not qualify as scientific: it makes no very exact statements, nor can very precise predictions be made from it. Yet not a scholar in the world would deny scientific status to the theory of evolution, even without its underpinnings in modern genetics. What makes a science are its aims, not its results. If it aims at establishing more or less general relationships between properties of nature, when the test of the truth of a relationship lies finally in the data themselves, and the data are not wholly manufactured—when nature, however stretched out on the rack, still has a chance to say "No!"—then the subject is a science. By these standards all the social sciences qualify—even history. The humanities do not. Much fiction, for instance, is very true to life, but the standard by which fiction is judged is certainly not in general this kind of truthfulness.

Much less often asked, though much more interesting, is the question: What sort of a science is social science? Yet I shall go wrong at the very beginning if I seem to imply in this question that social science differs radically from other science. The differences are matters not of kind but of degree. The enterprise of science faces everywhere the same characteristic problems. What forms do they take in the social sciences? How successful are the social sciences in coping with them, compared with one another and with the physical and biological sciences? If there are different degrees of success, what are the reasons for the differences? These are the questions I shall address myself to, without any hope of answering them fully.

I believe the questions are worth trying to answer, both to dampen the high hopes and to lighten the deep disillusion that sometimes afflict the students of the social sciences. I shall work on both the manic and the depressive phases of our collective psychosis. Briefly, you shall know the worst—and the worst shall make you free.

Though I sometimes teach history and have published papers in anthropology, I shall speak here primarily from the point of view of a sociologist, not only because that is by title my profession but also because sociology, it seems to me, suffers from the manic-depressive psychosis more severely than do the other social sciences. Its aims are larger, and its confidence smaller. In both directions it protests too much, and accordingly it stands in greater need of therapy.

If mine were questions to be answered by the philosophy of science, I should keep quiet, for the philosophy of science is well advanced. I shall need philosophy, but I shall need something more—the qualities of a student of comparative science—and I do not know that there is such a thing. If there were, how would he differ from a philosopher? Let me illustrate. Supposing explanation to be one of the aims of any science, a philosopher may hope to make clear to us what explanation is. In so doing he may well cite examples of the types of explanation developed in different sciences, and in this sense he will be comparative. Less often will he compare the sciences with respect to the practical difficulties they encounter in attaining to any explanation at all. It is just this latter sort of comparison that I shall need to make here.

Let me say at once that if the social sciences are in some ways less successful than the physical sciences, and some of the social sciences less successful than others, I do not—at least not in my sober moods—believe the reason to be that the scholars in the less successful fields are less intelligent, though success does tend to attract intelligence. And I never believe, though it is a common belief, the reason to be that the less successful fields are younger, and so have not had time to show what they can do. Sociology, for one, is not all that young—if it begins with Aristotle it is practically as old as physics—and in recent years it has been very energetic. No, I think the reason is of a different kind. An occasional thoughtless thinker asserts that science is a free creation of the human spirit. But how free? And where free? There is something intractable out there—call it the world, call it nature—intractable especially in not being all of a piece, all of the same grain, which may make it more difficult to create in some fields than in others. The difference finally lies neither in the mind nor in the subject-matter but in the relation between the two—the problems that the materials of the different sciences present for the mind trying to bring order out of their different kinds of chaos.

PROPOSITIONS

Any science has two main jobs to do: discovery and explanation. By the first we judge whether it is a science, by the second, how successful a science it is. Discovery is the job of stating and testing more or less general relationships between properties of nature. I call this discovery only because in many sciences the relationships were unknown before research revealed them: for instance, the discovery that bats navigate on the sonar principle. As we shall see, discovery in this sense, particularly discovery of the more general relationships, is much less characteristic of the social sciences than of the others, making one of the most striking differences between them.

A discovery takes the form of a statement of a relationship between properties of nature. Let us be sure we understand what this means. Take Boyle's familiar law: The volume of a gas in an enclosed space is inversely proportional to the pressure on it. A statement, a sentence like this, consists of two parts: first, a reference to what the relationship applies to—gas in an enclosed space—and second, a specification of the relationship between the properties, which must, of course, be at least two in number. Here the two properties are volume and pressure, and the relationship is inverse proportionality: if pressure goes up, volume will go down. Volume and pressure are continuous variables. In another variety of this kind of sentence, the properties, to speak loosely, can take only two values, as in the sentence: A man who loses his kidneys is dead. Here the variables are really classes: first, having kidneys or not having kidneys, and second, being alive or dead. And the relationship between the two is association: not having kidneys is definitely associated with being dead. Sentences of these two varieties I shall call "propositions." Propositions are the one essential product of any science.

In the words of Percy Bridgman, all propositions are

accompanied, implicitly or explicitly, by a "text." [1] In the case of Boyle's Law, the text would include answers to such questions as: What is a gas? What are pressure and temperature? How are they defined and measured? The text might also include a statement of the conditions within which the relationship held good. Boyle's Law holds good under the condition that the temperature of the gas is constant.

I have said that propositions, statements of relationships between properties of nature, were "more or less general." When I assert that the battle of Hastings was fought on October 14, 1066, I am certainly stating a relationship, but it is a relationship of association between a single event and a single time. If I asserted that all decisive battles were fought in October, the statement would, if true, begin to have some generality. And if I asserted that all battles whatsoever were fought in October, the generalization would be, in the terms used here, more general still. In the same way, Boyle's Law, which applies to all gases in an enclosed space and at constant temperature, is less general than a law applying to all gases at any temperature. But let us not worry much at the moment about the degree of generality of propositions. To have stated and tested a proposition of any degree of generality is no mean achievement. Let us remember Mr. Justice Holmes's dictum: "I always say that the chief end of man is to form general propositions," and let us not altogether

---

[1] P. W. Bridgman, *The Nature of Physical Theory* (Princeton, N.J.: Princeton University Press, 1936), pp. 59-61.

forget what he added: "And no generalization is worth a damn." [2]

NONOPERATING DEFINITIONS

I suppose every professor has horrid moments of feeling that he is teaching his students everything but what they really need to know, everything but the fundamentals. One reason why I have made the, after all, rather obvious points of the last few paragraphs is that I seldom teach my students how to recognize the different kinds of sentence that appear in the literature of social science, and I take the opportunity, belatedly and vicariously, of doing so now. Especially they need to be able to recognize a real proposition, or rather how to tell a real proposition from other kinds of sentence, for these nuggets are often few and far between. If, as Bridgman says, every proposition is accompanied by a text, the text in much of social science seems to take more room than it does in physical science. Indeed in some sociological writings no room is left for anything else.

Yet real propositions do appear in the literature of social science, and so do definitions of the terms that occur in them, the equivalents of the definition of

[2] M. DeW. Howe, ed., *Holmes-Pollock Letters* (Cambridge, Mass.: Harvard University Press, 1961), II, 13.

pressure that accompanies Boyle's Law. These I call "operating definitions," because we actually work with them. An example might be a definition of the term "frequency" to accompany the proposition: The more valuable a man perceives the result of his action to be, the more frequently he will perform the action. I want my students to be able to distinguish operating definitions and real propositions from two other kinds of sentence, similar in form to definitions and propositions respectively, which appear very often in the literature of social science, particularly in introductory texts and in "general theory." These I call "nonoperating definitions" and "orienting statements."

Examples of nonoperating definitions include the definitions of some so-called central concepts in sociology and anthropology, concepts the workers in these fields take to be the glories of their sciences. Thus a "role" is the behavior expected of a man occupying a particular social position. And a "culture" is the inherited pattern of living of the members of a society. These are nonoperating definitions because they do not define variables that appear in the testable propositions of social science. Though "roles" and "cultures" could each perhaps be analyzed into clusters of variables, they certainly are not such themselves. It would be absurd to say: "The more the role, the more the something else." We might indeed say: "The more specific the role, the lower the social position in which the behavior is expected." But here the variable would be specificity and not role itself.

This example suggests that "role" may have the status in sociological propositions that "gas" has in Boyle's Law: we might speak of the specificity *of* the role as we speak of the pressure *on* the gas. But I am not sure that the parallel holds. Certainly the status of the two is not exactly alike. For some propositions, like Boyle's Law, that hold good of gases do not always hold good of non-gases—liquids and solids—but it is far from clear that there are propositions that hold good of roles but not of non-roles (whatever they may be). That is, the word "gas" makes a difference in meaning, and "role" may not.

I think the same sort of thing is true of "culture." But here I add a comment that gets me a little ahead of my argument. An anthropologist friend once said to me, in pointing out the usefulness of this concept: "If someone asks me, for instance, why the Chinese do not like milk, I can only say, 'Because of the culture.' " [3] All I could say in turn was that, if that was all *he* could say, he was not saying much. All that the use of the word "culture" implied was that disliking milk had been characteristic of the behavior of some Chinese for some generations. But we knew that already; "culture" did not add anything. What we should have liked to know was why milk, specifically, rather than, say, tea was disliked. Talking about culture did not

[3] For the notion that the concept of culture explains something, see, especially, Clyde Kluckhohn, *Mirror for Man* (New York: McGraw-Hill, 1949), pp. 17-44.

answer this question at all—not at all. More generally, "explanation by concept" is not explanation.

Yet I am loath to argue that the concepts "role" and "culture" are useless. What I want to be sure of is that we recognize the sort of usefulness they possess. They tell us roughly the kinds of thing we are going to talk about. They and their definitions tell us that we are going to talk about expected behavior and inherited patterns of behavior, and it may indeed be well for a new student to be forewarned. But sooner or later we must stop "being about to" talk about something and actually say something—that is, state propositions. Lingering over nonoperating definitions may actually get in the way of this primary job of science. This happens, I think, when nonoperating definitions are multiplied and elaborated into a nonoperating conceptual scheme (called a "general theory"), as in much—not all—of the work of Talcott Parsons. Some students get so much intellectual security out of such a scheme, because it allows them to give names to, and to pigeonhole, almost any social phenomenon, that they are hesitant to embark on the dangerous waters of actually saying something about the relations between the phenomena—because then they must actually take the risk of being found wrong. The failure to state real propositions leads in turn to a failure to create real theories, for, as we shall see, a real theory consists precisely of propositions. I sometimes think we need not be at such pains introducing our students to social science. Start them out at once with real propositions. They would find out soon enough what we were going to talk about: we should already be talking about it.

ORIENTING STATEMENTS

Just as "role" and "culture" are famous concepts, so what I call "orienting statements" include some of the most famous statements of social science. One is Marx's statement that the organization of the means of production determines the other features of a society. This is more than a definition and resembles a proposition in that it relates two phenomena to one another. But these phenomena—the means of production and the other features of a society—are not single variables. At best they are whole clusters of undefined variables. And the relationship between the phenomena is unspecified, except that the main direction of causation—determination—is from the former to the latter. Whereas Boyle's Law says that, if pressure goes up, volume will assuredly go down, what Marx's Law says is that, if there is some, any, change in the means of production, there will be some unspecified change or changes in the other features of society. Put the matter another way: Boyle will allow one to predict *what* will happen; Marx will only allow one to predict that *something* will happen. Accordingly I cannot grant his law the status of a real proposition.

In taking Marx's statement thus out of context, I do not in the least mean to imply that this is all he had to say about the relations between the infrastructure and the superstructure of society, or that his writings do not include other statements that are real propositions, or that this particular statement is unimportant. That is far from my view.

Another example of an orienting statement is the assertion by Parsons and Shils that, in social interaction between any two persons, the actions of each are sanctioned by the actions of the other.[4] This is an important statement in that, in my view, the beginning of wisdom in the study of social behavior is to look at it as an exchange between at least two persons, in which the action of each rewards or punishes—that is, sanctions—the action of the other. But the statement in itself does not say what effect a change in the behavior of one will have on the behavior of another. Like Marx's Law, it implies that there will be *some* effect, but does not begin to say what. Only if Parsons and Shils had gone on to say, for instance, that the more rewarding (valuable) to one man is the action of the other, the more often will the first perform the action that gets him the reward—only then would they have stated a real proposition. Much of what they say suggests that they believe this proposition to be true, but they manage to avoid coming right out with it.

[4] T. Parsons and E. A. Shils, eds., *Toward a General Theory of Action* (Cambridge, Mass.: Harvard University Press, 1951), pp. 14-16.

For my third example I do not take an isolated statement but a passage from a book, chosen for no better reason than that I read it recently. If it were a bad book, that fact would get in the way of my making the point that passages more or less like this one have been appearing for many years and in immense numbers in all the best literature of social science:

An individual is born into a social system that possesses a culture. The socialization of that individual is a threefold process. It involves the inculcation of the culture upon the individual by the social system. The transmission of culture through socialization is never complete. The individual learns only a selected number of elements in the culture of his society. He introjects, or commits himself to, even fewer elements of that culture. In so doing he brings to bear the influence of his own personality upon the survival and growth of the culture. Also, the social system itself does not act upon the individual in his socialization. It is individual members of the system who act upon him. Their influence upon him reflects the uniformity of relationships that comprise the social system. Their influence also reflects their own idiosyncratic response to culture, based on the dynamics of their personalities. In this way, the individual is trained in the common bonds of society. But the pattern of his being taught and the lessons he learns are unique to him.[5]

[5] R. C. Hodgson, D. J. Levinson, A. Zaleznik, *The Executive*

In a way, this is all sound as a bell. I suppose we could find out, for instance, what words like "culture" and "socialization" meant, and would even agree that culture was transmitted through socialization. But let us ask ourselves this question: Where in the passage do we find a single statement from which we could tell what specific change would occur, or even probably occur, along any one dimension of human behavior if there were a specific change along another? Yet it is the business of a science to make such statements. The passage tells us that things like culture, socialization, and the social system are all important and all somehow related to one another, but it tells us nothing *about* them. It is all true—and all powerless. After so many years of orientation, do we and our students really need so many weak truths?

Much writing in social science consists of orienting statements when it does not consist of nonoperating definitions. Orienting statements do not qualify as real propositions: they are of little use in prediction and of none at all, as we shall see, in explanation. Yet I should be slow to argue that they did no good in other ways. I must testify, perhaps complacently, that I personally have been greatly helped both by Marx and by Parsons and Shils. I claim that statements of this sort are really imperatives, telling us what we ought to look into further or how we ought to look at it. This is the reason why I call them orienting statements. Look at the rela-

Role Constellation (Boston: Harvard Graduate School of Business Administration, 1965), p. 37.

tions between the means of production and the other features of society, for if you look, you will surely find! Look on social behavior as an exchange, for then you will begin to make progress! And, God knows, with the help of Marx at least, scholars have made progress. Looking where he pointed, they have discovered and tested statements that, if of smaller scope than Marx's, still have more of the character of real propositions.

Yet the very success of Marx's Law in being useful in its particular way teaches us that we should not mistake orienting statements for either the empirical or the theoretical results of science. A statement that tells us what to study or how to study it is an important statement. But it tells us little about the thing studied. In Merton's words, it gives us an approach, not an arrival.[6] Let us not exhaust ourselves in the preliminaries, lest we fail at the consummation. We are always getting around to saying something we never actually come out with. But sooner or later a science must actually stick its neck out and say something definite. If there is a change in x, what sort of change will occur in y? Don't just tell me there will be *some* change. Tell me *what* change. Stand and deliver!

THE FINDINGS OF SOCIAL SCIENCE

And the social sciences do this. Though nonoperating definitions and orienting statements are comparatively

[6] R. K. Merton, *Social Theory and Social Structure*, rev. ed. (Glencoe, Ill.: The Free Press, 1957), p. 9.

prevalent in them, especially in anthropology, sociology, and political science, and are comparatively often mistaken for real definitions and propositions, yet the social sciences now have a very large number of solid findings to their credit. At the turn of the century the mathematician Poincaré could sneer that ". . . sociology is the science that possesses the most in the way of methods and the least in the way of results." [7] He could not fairly say so now. Though we still talk endlessly about methodology, the other side of the balance has been redressed. Anyone, for instance, who reads the useful book by Bernard Berelson and Gary A. Steiner, *Human Behavior: An Inventory of Scientific Findings*,[8] should be impressed with the number of generalizations (propositions) in this field that have now been pretty well tested against data.

Choosing almost at random, let us get an idea of their variety within some sub-field, such as social stratification.[9] Every society, certainly every society of any size, is stratified by class or status. The rate of intergenerational mobility between classes is currently about the same in all highly industrialized nations. Family instability (divorce, separation, and abandonment) is greatest in the lower class, next in the upper, and least in the middle. The higher the class, the later the average age at marriage. And so forth. If the first job of a

[7] H. Poincaré, *Science et Méthode* (Paris: Flammarion, 1909), pp. 12-13.
[8] New York: Harcourt, Brace & World, 1964.
[9] *Ibid.*, pp. 453-91.

science is to establish generalizations, social science has established a great many.

But look at the characteristics of these propositions. Except for the first—that all societies are stratified—all of them state only central tendencies. It is not true of all members of an upper class that they marry late, but only of the average. And when the propositions state relationships between variables, the nature of the relationship, the function, is not very specific. We know that a rise in class position means a rise in the age of marriage; what we cannot say is, for instance, that one increases as the logarithm of the other. Sometimes a social science can get a little closer to specifying the shape of the function, as in the so-called law of diminishing marginal utility in economics: the curve relating the quantity of a good received by a man and the value to him of a unit of the good is concave downward. But few of our propositions ever state the exact function—which is one of the reasons why our science is not an exact science. Still, they are likely to say at least this much: that, for instance, as the value of one of the variables increases, the value of the other increases too—which is enough, just enough, to make them real propositions and not simply orienting statements.

Even more important, our propositions, though they are generalizations all right, are seldom very general generalizations. They are known to hold good only within rather narrow limits, only within western industrial societies, for instance. Or if the limits are not known, they are still shrewdly suspected of being narrow. And finally, with one class of exceptions, which I shall speak of much later, even our apparently most general generalizations, like the proposition that all societies are stratified, do not possess much explanatory power. Thinking of these, I have sometimes entertained the hypothesis that in social science the greater the generalization, the less its explanatory power. But explanation brings in a new kind of consideration.

THE NATURE OF EXPLANATION

Most people interested in comparing the social sciences with the natural sciences, especially those interested in making sure that social science *is* a natural science, emphasize the greater difficulty the social sciences face in establishing, against data, the empirical truth of its propositions. It is certainly less easy in the social sciences than in some physical and biological sciences to manipulate variables experimentally and to control the other variables entering into a concrete phenomenon, so that the relationship between those the scientist is interested in at the moment shall be, beyond question, unmasked and stand out clearly. It is less easy to control the variables because it is less easy to control men than things. Indeed it is often immoral to try to control them: men are not to be submitted to the indignities to which we submit, as a matter of course, things and animals. Hence the relative prominence in some of the social sciences, even increasingly in history, of other methods of controlling variables, methods thought somehow to be less satisfactory, such as the use of statistical techniques.

I shall have no more to say about this difference between the social sciences and the others. Admittedly it is important, but it is also rather well understood, and much intelligence of a high order has been devoted to finding methods of dealing with the problem. Moreover, some of the biological sciences, such as medicine, suffer from difficulties of control almost as much as do the social sciences. Much less well understood are the differences between the social and the other sciences in the matter of explanation.

Though stating and testing relationships between properties of nature is what makes a science, it is certainly not the only thing a science tries to do. Indeed we judge not the existence, but the success, of a science by its capacity to explain. If there is one thing I should like my students to learn but seldom teach them is what an explanation is—not that it his hard to do. Again, no "big" word is more often used in social science than the word "theory." Yet how seldom do we ask our students—or, more significantly, ourselves— what a theory is. But a theory of a phenomenon is an explanation of the phenomenon, and nothing that is not an explanation is worthy of the name of theory.

I am, of course, using "explanation" in the special sense of explaining why under given conditions a particular phenomenon occurs and not in one of the vaguer senses in which we use the word, as when we "explain" how to drive a car by telling a youngster what to do with the controls in various circumstances. In the special sense, the explanation of a finding, whether a generalization or a proposition about a single event, is the process of showing that the finding follows as a logical conclusion, as a deduction, from one or more general propositions under specified given conditions.[10] Thus we explain the familiar finding that there are two low and two high tides a day (actually a little longer than twenty-four hours) by showing that it follows logically from the law of gravitation under the given conditions that the earth is largely covered with water, that it rotates on its axis, and that the moon moves in orbit around it.

But let me go into more detail, using a humble example but one that in the past had good reason to interest me. As a boy swimming in the fundamentally rather chilly waters of Massachusetts Bay in summer, I discovered, as others had done before me, that for comfort in swimming, the water near the shore was apt to be warmer when the wind was blowing onshore

---

[10] The view of explanation adopted here is, I think, that of R. B. Braithwaite, *Scientific Explanation* (Cambridge: Cambridge University Press, 1953) and of C. H. Hempel, *Aspects of Scientific Explanation* (New York: The Free Press, 1965), pp. 229-489. Philosophers will note that I have dodged the issue of the implicit definition of "theoretical" terms. See also G. C. Homans, "Contemporary Theory in Sociology" in R. E. L. Faris, ed., *Handbook of Modern Sociology* (Chicago: Rand-McNally, 1964), pp. 951-77.

—towards the shore—than when it was blowing off-shore. By thoroughly unsystematic statistical methods I tested the discovery and found it true. But why should it be true? I shall try to give the essentials of what I believe to be the correct, though obvious, explanation, without spelling it out in all its logical, but boring, rigor.

Warm water tends to rise. The sun warms the surface water more than the depths. For both reasons, surface water tends to be warmer than deeper water. The wind acts more on the surface water than it does on the depths, displacing it in the direction of the wind. Accordingly an onshore wind tends to pile up the warmer water along the shore, while an offshore wind tends to move it away from the shore, where, by the principle that "water seeks its own level," it is continuously replaced by other water, which, since it can only come from the depths, must be relatively cold. Therefore water along the shore tends to be warmer when the wind is blowing onshore than when it is blowing offshore. Q.E.D.

Simple though it is, the characteristics of this explanation are those of all explanations. Each step of the argument is itself a proposition stating a relationship between properties of nature: between, for instance, the temperature of water and the direction of its movement, up or down. That is why propositions are so important. Some of the propositions are more general than others. In the example, some of the more general propositions are that warm water tends to rise and that water seeks its own level. They are more general in that they apply to all water and not just water along a coast. Some of the propositions state the effect of the given conditions, such as that the wind sometimes blows onshore and sometimes offshore. By calling them given conditions we mean simply that we do not choose to explain them in turn—we do not choose to explain why the wind sometimes blows onshore—though no doubt we could do so. And the proposition to be explained, the *explicandum*—in this case the difference in temperature of coastal water under onshore and offshore winds—is explained in the sense that it follows as a matter of logic from the general propositions under the specified given conditions. That is, the *explicandum* is deduced from, derived from, the other propositions, the whole set forming a "deductive system." The reason why orienting statements cannot play a part in explanation is that little in logic can be deduced from them.

Note that, if the *explicandum* can be deduced from the general propositions under the given conditions, the general propositions cannot be deduced in turn from the others in the set, any more than in the classic syllogism we can deduce that all men are mortal from the facts that Socrates is a man and that Socrates is mortal. That is, the process of deduction runs in one direction but not the other in the set of propositions. If it did both, the argument would be circular. On the other hand, the general propositions in our example can themselves be explained by, can themselves become the *explicanda* of, other deductive systems containing still more general propositions. That hot water rises is ultimately explained by propositions of thermodynamics relating the temperature of any substance to its volume and thus to its weight per unit volume. That water seeks its own level is ultimately explained by the law of gravitation. But as we move towards more and more general propositions, we reach, at any given time in the history of science, propositions that cannot themselves be explained. If we can judge from experience, this condition, for any particular proposition, is unlikely to last forever. Newton's law of gravitation stood unexplained for some two hundred years, but can now be shown to follow from Einstein's theory of relativity. Nevertheless at any given time there are always at least a few unexplainable propositions.

The explanation of the relation of water temperature to wind direction is also the theory of this phenomenon. But of course scientists generally use the word "theory" in a broader sense than this. They use it to refer, not just to an explanation of a single phenomenon, but to a cluster of explanations of related phenomena, when the explanations, the deductive systems, share some of the same general propositions. Thus someone might write a book called *The Theory of Water Temperatures*, which might explain the relations between variations in temperature and a number of other conditions besides the one chosen in our example, and which would apply, in doing so, a number of the same general propositions from thermodynamics and mechanics. Naturally any scholar is free to use the word "theory" in any way he likes, even for something different from what I call theory, provided he makes clear just how he is using it and does not, by slurring over the issue, claim for his kind of theory, by implication, virtues that belong to a different kind. All I submit here is that, normally in science, "theory" refers to the sort of thing I have described.

If we like, we can look on theory as a game. The winner is the man who can deduce the largest variety of empirical findings from the smallest number of general propositions, with the help of a variety of given conditions. Not everyone need get into the game. A man can be an admirable scientist and stick to empirical discovery, but most scientists do find themselves playing it sooner or later. It is fascinating in itself, and it has a useful ulterior result. A science whose practitioners have been good at playing it has achieved a great economy of thought. No longer does it face just one damn finding after another. It has acquired an organization, a structure. When Newtonian mechanics reached this sort of achievement it became the first thoroughly successful science, and other sciences have since become successful in the same way. But if theory is a game, it must like other games be played according to the rules, and the basic rules are that a player must state real propositions and make real deductions. Otherwise, no theory!

EXPLANATION IN SOCIAL SCIENCE

There are scholars who argue that, if social science is a science at all, it is a radically different kind of science from the others, and that it makes a mistake pretending to be the same sort of thing. I do not believe this in the least. The content of the propositions and explanations is naturally different in social science, because the subject matter is different, from what it is in the others, but the requirements for a proposition and an explanation are the same for both. And so long as the compulsion to be scientists does not rob us of our native wit and prevent our seeing what is there in nature to be seen, I believe that the social sciences should become more like the others rather than less. As we have come to accept, with all the difficulties that acceptance entails in our case, the standards of natural science for testing the truth of propositions, so we should take more seriously the standards of natural science in explanation. In that we have been laggard. Though the social sciences face special difficulties with explanation as they do with testing propositions, we can still do better than we have done.

It is not in its findings, which are now numerous and well attested, that social science gets into trouble, but in its explanations. The trouble takes somewhat different forms in different fields, but explanation, theory, is always its seat. Let me briefly take up some examples.

Most scholars would recognize that economics is the most advanced of the social sciences. It certainly possesses real theories, both in the micro- and the macro-fields. The question with economics is how general its theories are. The so-called laws of supply and demand are certainly not general. The demand for perfume, for instance, does not obey the law: the higher the price of a perfume, the greater the demand for it, at least up to a point. The question for economics is: What are the more general propositions from which, under different conditions, both the agreements with, and the exceptions to, economic laws may be deduced? Economists have, I think, acknowledged the pertinence of the question, but they have not reached agreement on an answer, perhaps because they have been successful enough in other ways not to feel the need of one.

History is at the opposite pole. It possesses an enormous range of empirical findings, findings, that is, of a rather low order of generality. It certainly claims to explain, but it pretends—or most historians pretend—to have no theories. A theory ought to include general propositions. The historians have looked for general propositions in their subject-matter, found none that they recognized as such, and concluded that they had no theories. I think they looked in the wrong place— but that is not quite fair. What they did was miss, as many of us do, the object hidden in plain sight. I think history does have general propositions, but history does not mention them, leaves them unstated.

If history has many explanations and no theories,

sociology—and anthropology resembles it—sometimes appears to have many theories and no explanations. Certainly there are sociologists who claim to have very general theories, general enough indeed to encompass all the other social sciences. But when examined closely the theories often fail as explanations. They may consist of a matrix of definitions, and of nonoperating definitions at that. And when the theories try to state relationships between the properties defined, the statements may turn out to be orienting statements and not real propositions. On both counts they fail to qualify as deductive systems.

Besides theories, sociology possesses, as I have pointed out, a great many tested propositions. Again, most of them are of a low order of generality, and so cry out for the kind of organization that a good theory could provide. At this very point official sociological theory fails them. There is no science in which the rich and varied findings bear so little relation to the theories, in spite of endless pleas that they ought to be related. Indeed the faults of the theories, as theories, seem positively to get in the way of the organization of the findings. The theorists are always about to get into contact with the data, but never do; and the empirical researchers, while waiting in vain for help from on high, do not create their own theories because "theory" is a special field pre-empted by others.

Sociology and anthropology even possess some very general generalizations, the so-called "cultural universals," such as that all societies have incest taboos or, the one mentioned earlier, that all societies are stratified. Unfortunately it is not enough, in order to be useful in a theory, for a proposition to be general. It must also have explanatory power, and the cultural universals do not. From them alone one can derive only one kind of empirical proposition—propositions about single instances. These propositions refer to societies, and a society for them is a single instance. So if Xia is a society, it will be stratified—that is all one can say. This is a much lower degree of explanatory power than that possessed, for instance, by Newton's laws, from which a wide variety of propositions can be derived, not only propositions about individual instances, but also many propositions, like the one about the tides, that are themselves generalizations. Far from helping us explain anything, propositions like the one about stratification themselves demand explanation. Why indeed should all societies be stratified?

The characteristic problems of social science, compared with other sciences, are problems of explanation. Explanation is the deduction of empirical propositions from more general ones. Accordingly, in the matter of explanation, the problems of social science are two in number. What are its general propositions? And can empirical propositions be reliably deduced from them? For it is conceivable that, even if a science possesses general propositions, it may not be able to do much with them in the way of deducing the empirical propositions it most wishes to explain. To these two problems, in this order, I shall now turn.

2

GENERAL
PROPOSITIONS

The characteristic problems of social science are problems of explanation, and the first of them is, What are its general propositions? Before the rise of academic anthropology and sociology at the end of the nineteenth century, the nature of the answer would have seemed obvious to most scholars. They would have answered: propositions about "human nature," about the psychic characteristics men share as members of a single species. It was anthropology and sociology, as we shall see, that first raised the possibility of there being an alternative answer. But at the very time the alternative was being put forward, the work that would furnish a reply to it was in progress—work in modern psychology. My contention will be that the original answer was correct, provided we accept the view modern psychology takes as to the essentials of human nature. It is not really a new view, except that it eliminates from the list of essentials some things that might have been considered such in the past. In this book we shall be apt to find ourselves coming back to what look like old-fashioned ideas by way of new-fangled research.

BEHAVIORAL PSYCHOLOGY

In the last chapter I suggested that we look at the book by Berelson and Steiner, *Human Behavior: An Inventory of Scientific Findings*. In it we found a large number of tested propositions, but we also saw that many of them possessed either little generality or little explanatory power. If we now look through the book again, we shall find, near the beginning, propositions of a somewhat different kind. These are the propositions included in the chapter called "Learning and Thinking." [1]

One such proposition is: "When a response is followed by a reward (or 'reinforcement'), the frequency

[1] B. Berelson and G. Steiner, *Human Behavior* (New York: Harcourt, Brace & World, 1964), pp. 133-237.

or probability of its recurrence increases." This I call a proposition about the effect of the *success* of a person's action on its recurrence. Several other propositions in the same field ought to have been included. Thus, if a response (action) was followed by a reward under particular conditions (stimuli) in the past, the reappearance of similar conditions makes more frequent or probable the recurrence of the action. This I call a proposition about the effect on a person's action of his *perception* of the situation attending his action. Again, the higher the *value* a person sets on the reward, the more likely he is to take the action or repeat it; and the value of the reward increases the more the person has been *deprived* of it, as in hunger, and decreases the more nearly he has been *satiated* with it. If the reward has a negative value, that is, if the action has been punished, the probability of the action's recurring decreases. Several propositions about emotional behavior ought also to be included, such as the familiar one that, if a person is *frustrated*, he is apt to take action that may be described as aggressive.[2]

Scholars in the field of "learning and thinking" would not agree as to just what propositions should be included on a list of fundamental ones. Nor, since like all scholars they set a high value on their own special terminologies, would they agree on the wording of the propositions. But most of them would agree that some

[2] For a fuller discussion of these propositions see G. C. Homans, *Social Behavior* (New York: Harcourt, Brace & World, 1961), pp. 30-82.

of the propositions on the above list should be included, and most of them would agree on their substance.

Propositions like these constitute what is often called "learning theory," but the name is misleading, for they continue to hold good of human responses long after the responses have, in every usual sense of the word, been learned. Though a man may learn quickly that he can buy food cheaply at a supermarket, his behavior does not for that reason cease to be governed by the value-proposition, and he will go back to the supermarket whenever he needs to buy food. Accordingly I prefer to speak of behavioral psychology rather than learning theory.

These propositions refer to voluntary behavior. A more complete set should include propositions about involuntary actions, the reflexes, such as the familiar knee-jerk. If I say no more about the reflexes, the reason is that they do not include the features of behavior of most interest to social scientists.

What has been called the "rational-choice model" of human behavior coincides in part with the body of propositions of behavioral psychology. The coincidence has not always been recognized, because the rational theory has usually been put forward not by psychologists but by other scholars, such as economists and mathematicians interested in explaining the process of decision-making. The main proposition of the rational theory in one of its forms may be stated as follows: In choosing between alternative courses of action, a person will choose the one for which, as perceived by him, the mathematical value of $p \times v$ is the greater, where p is the probability that the action will be successful in getting a given reward and v is the value to the person of that reward. The effect of value on action in the rational theory is embodied in the value-proposition of behavioral psychology; the effect of the probability of success, in the success-proposition. But the rational theory is obviously more limited than behavioral psychology. Though it recognizes the importance of perception, and assumes that a man is rational in acting in accordance with his perceptions even though in the eyes of persons better informed his perceptions may be incorrect, it simply takes the perceptions as given and does not tie them back, as behavioral psychology does, to past experience. In the same way, it takes a person's values as given and does not tie them back to deprivation and to the process by which new values are learned. Nor is it at all clear how the rational theory would deal with emotional behavior. But we need not worry about the limitations of the rational theory, so long as we recognize that it is not an alternative to behavioral psychology; the two are in fact largely the same. This, I argue, is a first point in favor of the view that the social sciences share the same body of general propositions.

I cannot embark here on a treatise on behavioral psychology and spell out all the implications of the propositions. All I want to do is emphasize their main characteristics, and, above all, that they are psychological. They are psychological in two senses. First, they are usually stated and empirically tested by persons who call themselves psychologists. Though they are, as we shall see, used in explanation by all the social sciences, the field of research in which they lie is the special concern of only one of these sciences, psychology. Second, they are propositions about the behavior of individual human beings, rather than propositions about groups or societies as such; and the behavior of men, as men, is generally considered the province of psychology.

The propositions are not new, in the sense that they were once unknown and have had to be discovered. Though the language in which psychologists state them is unfamiliar because it aims at precision, their content is not. When we know what they mean, they do not surprise us, though some of their further implications may do so. Nor, we may guess, would they have surprised Cro-Magnon men. Indeed they are part of the traditional psychology of common sense.

Another part of that psychology this body of propositions leaves out or, better, de-emphasizes. The traditional psychology tended to argue that "human nature was the same the world over" in the sense that all human beings, or all of certain main kinds, shared certain very specific values. Thus all women were supposed to feel for their children something called "mother love"; they were supposed to find the care of their children naturally rewarding, and this was supposed to account, for instance, for universal features of the family as an institution. Human nature is certainly not concretely the same the world over,

and modern anthropology and psychiatry have forced
us to recognize many surprising ways in which it
differs from society to society, from group to group,
and from individual to individual. Faced with facts
like these, the new psychology has not abandoned
the idea of the universality of human nature, but
has given up specific similarities for more general
ones. It tends to emphasize, not that men hold similar values, but that, whatever their values may be,
these values have similar effects on their behavior. A
particular kind of reward may be valuable to one man
or to the members of one group, a different kind valuable to another; and since the pursuit of different rewards often requires different actions, what the two
men or groups do may differ concretely. Yet the proposition: the more valuable the reward, the more frequent
or probable the action that gets the reward, holds good
for both. Even if many men and women do share in
some degree values built into them genetically and
biologically—even if in this sense there is indeed something to be called "mother love"—still the actions they
can learn to adopt in attaining these values are very
varied: the actions are specified by instinct to a far
lower degree than was assumed by the traditional psychology. There are also values, like that of money, that
are not genetically determined but are themselves
acquired by learning; different persons in different
circumstances may acquire different ones, and yet the
process is the same for all: like money, they become
values in themselves by proving to be the means to

other, more primordial values.[3] It is often said that the new psychology emphasizes the plasticity of human behavior. It has kept—and extended and tested—just that part of the old psychology which, stressing how similar men are in *how* they learn, accounts paradoxically for how different they can be in *what* they learn.

The propositions of behavioral psychology have high generality. They purport, at least, to hold good of all human beings and indeed of all the higher animals, modified in the case of the animals by mechanisms called instinctual that are rather more specific in their operation than such mechanisms are in men. But we have learned that, for our purposes, it is not enough for propositions to be general: they must also possess explanatory power.

In arguing that they do command wide explanatory power, I might call upon everyday experience. Even the social scientists who are most opposed doctrinally to psychological explanations can be heard to use these propositions in some vulgar form to explain much of their own behavior and that of their fellows, as individuals or as groups. They use them in the real world, where having a good theory really counts. In their books they leave the propositions out while they are up on the high horse of "theory" but sneak them back in, without admitting or even recognizing what they are doing, when they finally have to get down to ex-

[3] See especially A. W. Staats and C. K. Staats, *Complex Human Behavior* (New York: Holt, Rinehart and Winston, 1963), pp. 48-54.

plaining something.[4] In fact, their actual explanations *are* their actual theories. But it is difficult to make this kind of evidence tell, as many men have no trouble denying their own experience.

I believe that this body of propositions—though it is the academic property of one of the social sciences, psychology—provides for all of them at present the most general propositions used in explanation. Not only in their subject matter, human behavior, but in their general statements about human behavior, the social sciences are one science. I cannot hope to prove that this is so, for that would require my setting forth all the explanations in question. But I may hope to persuade, and I shall try to do so by illustrating the use of some of the propositions for explanatory purposes in a variety of social sciences.

PSYCHOLOGICAL EXPLANATION IN HISTORY

Take history first. In recent years there has been controversy over the nature of explanation in history. In the course of the controversy the philosopher Hempel took the position, which I share, that an explanation would include general propositions or, as he called them, laws. Whereupon another philosopher, Scriven, wrote: "Suppose we wish to explain why William the

[4] See G. C. Homans, "Bringing Men Back In," *American Sociological Review*, XXIX (1964), 809-18.

Conqueror never invaded Scotland. The answer, as usually given, is simple enough; he had no desire for the lands of the Scottish nobles, and he secured his northern borders by defeating Malcolm, king of Scotland, in battle and exacting homage. There appear to be no laws involved in this explanation." [5]

There are indeed no laws in this explanation, but is it by our standards an explanation at all? With the use only of the propositions explicitly stated, no deductive system can be constructed. Yet if a major premise, which is now lacking, were supplied, an explanation would emerge. In sketch form, it would run like this:

1) The greater the value of a reward to a person, the more likely he is to take action to get that reward.

2) In the given circumstances, William the Conqueror (a particular person) did not find the conquest of Scotland at all valuable.

3) Therefore he was unlikely to take action that would win him Scotland. [6]

---

[5] M. Scriven, "Truisms as the Grounds for Historical Explanations" in P. Gardiner, ed., *Theories of History* (Glencoe, Ill.: The Free Press, 1959), pp. 443-75. I have used this example in G. C. Homans, "Contemporary Theory in Sociology" in R. E. L. Faris, ed., *Handbook of Modern Sociology* (Chicago: Rand-McNally, 1964), pp. 951-77. For a fuller account of William's action see D. C. Douglas, *William the Conqueror* (Berkeley: University of California Press, 1964), pp. 225-28.

[6] Technically, this does not explain why William *did not* take the action in question but only why he was *unlikely* to do so. The reason is that the major premise is probabilistic.

One can get pretty obvious if one tries, and in a good cause I believe in trying. There is an important point to be made here, and it is this: the alleged explanation, which Scriven said contained no law, can readily be turned into a real explanation if an unstated major premise is supplied, the major premise being Proposition 1 in the admittedly sketchy deductive system. And this proposition, which is very general, turns out to be the value-proposition of behavioral psychology.

Note, by the way, that the explanation would become circular if we had no evidence, other than that William did not in fact conquer Scotland, for his not finding the conquest of that country rewarding. In some, but far from all, historical explanations we do have some kind of independent evidence about the values of the actors.

What is true of Scriven's explanation is true of most historical explanations: the major premises are left unstated—in the language of logic the explanations are "enthymemic" [7]—but if the missing premises are supplied they often turn out to be propositions of behavioral psychology. According to what is to be explained, they may also be propositions of other sciences, such as medicine. For instance, an unstated major premise in some historical explanations is that a man whose head is severed from his body is dead. But I am sure that propositions of behavioral psychology are the ones that need to be supplied most often.

[7] See M. White, *Foundations of Historical Knowledge* (New York: Harper & Row, 1965), pp. 57-59.

I am far from arguing that the general propositions of history should not be left unstated. Most historical explanations take part in a series of explanations forming a genetic chain,[8] in which the given conditions of one deductive system become the *explicanda* of others. Often the actions of one man (or group) are explained by a deductive system that takes the previous actions of other men as given. But the actions of the others may in turn be explainable by another deductive system, taking now the previous actions of third parties as given, and so on. The individual deductive systems, thus linked together, will be apt to contain, since they are all concerned with the actions of men, some of the same general psychological propositions. History may not repeat itself, but the general propositions of history certainly do—or would do so if the genetic chains were fully spelled out. But their repetition would be boring to the reader and would take up much space in print. Moreover, the general propositions might well seem obvious. Indeed, Scriven's article is entitled "Truisms as the Grounds for Historical Explanations." Finally, the really difficult thing to explain is not what followed from the fact that William the Conqueror did not find the conquest of Scotland rewarding but why he did not find it so.

Fear of being obvious and repetitious and of diverting attention from crucial issues are good reasons for

[8] On genetic explanations, see C. G. Hempel, *Aspects of Scientific Explanation* (New York: The Free Press, 1965), pp. 447-53.

not always spelling out the major premises of history—so long as historians, in adopting this sensible practice, recognize what they are doing and do not let it fool them into believing that there are no generalizations in history as a science. Many historians have looked for sweeping generalizations about historical processes as such. They have found none of their own, and they have been sceptical of those put forward by their colleagues. I think they have not found them because they have been looking in the wrong place. The generalizations have been in plain sight all the time. The greatest generalizations of any science are those it in fact uses, whether it mentions them or not, as its major premises in explanation. The greatest generalizations in history are not propositions about the processes of history but about the actors in history, that is, men.

PSYCHOLOGICAL EXPLANATION IN ECONOMICS

Let us now turn to an example from economics, and to the explanation, not of a single event like William the Conqueror's retreat from Scotland, but of what are themselves propositions of some degree of generality. Micro-economics—so called to distinguish it from macro-economics, the study of large-scale economic systems—begins by relating the price of a commodity in some market to the quantities of the commodity bought and sold there, two of its propositions being the celebrated "laws" of supply and demand. Let us not worry about whether they are properly called laws: they, and inferences from them, certainly hold good in a number of different circumstances. The law of supply is often stated as follows: the higher the price of a commodity, the more of it a supplier will sell. I prefer to put it in probabilistic form: the higher the price of a commodity, the more likely is a supplier to sell some of it; and if he is likely to sell some, he is also likely to sell more—if he has it. Now if the price of a commodity is a measure of the value to the supplier of the result of selling a unit of the commodity—the higher the price the greater the value—then the law of supply follows directly from the general proposition used to explain the behavior of William the Conqueror: the greater the value of a reward to a person, the more likely he is to take action to get that reward. Note that in micro-economics the success of the action is not in question: it is assumed that a supplier can sell if he wants to, that buyers are on hand in the market.

As for the "law" of demand, I shall state it in the form: the higher the price of a commodity, the less likely is a buyer to buy some of it. To explain this law we have only to extend the same general proposition, as perhaps we ought to have done at the outset, by taking account of relative values: the greater the value of a reward to a person, relative to the cost of getting the reward, the more likely he is to take action to get it, where the cost is the value of the result of an alternative action forgone in performing the first. In the law of demand the price of the commodity is a measure of the value the purchaser forgoes when he spends his money on the commodity instead of saving it or spending it on something else. If the price goes up, the cost goes up and the net value of buying goes down, so that by the general proposition, as extended, the purchaser is less likely to buy.

It has long been known that the law of demand does not always hold good. I believe it does not always hold good of perfume and other commodities that go into what Veblen called "conspicuous consumption." Up to some point, the higher the price of a perfume, the more likely a consumer is to buy it. But even the exception can be explained by the same general proposition, for to a consumer of perfume, I am told, a rise in its price, provided the perfume and its high price are well known, positively enhances its value as a symbol of status, so that as the price increases, it is not just the cost that increases but the value too, and for a time the latter may increase disproportionately, making a consumer more rather than less likely to buy.

That the laws of supply and demand can be stated at all in elementary economics depends on certain institutional conditions. It depends on the existence of money as a standard in which prices can be expressed and on the existence of markets, whether the old market-place or its modern equivalents. It is much more difficult to explain the institutional conditions themselves than to explain behavior once the institutions are given. Elementary economics was lucky in being able to take institutions pretty much for granted. It could leave the harder jobs of explanation to the other social sciences, and to this it may owe the fact that it has progressed more rapidly than they have.

PSYCHOLOGICAL EXPLANATION OF CONFORMITY

Institutions are among the main concerns of anthropology, sociology, political science, and history. Their defining characteristic is a set of rules or, as sociologists call them, norms: statements specifying how persons ought, or ought not, to behave in particular circumstances. Thus money as an institution is accompanied by rules against counterfeiting coins and bills, and the market, by rules against one form or another of rigging the market, of conspiracies in restraint of trade. One can make much progress in anthropology and sociology by working directly with the institutions, showing how they are related to one another within a particular society or across many societies, without ever worrying about whether the institutional rules are obeyed. But sooner or later one must ask why the rules are obeyed at all; one must raise the question of conformity to norms. For if its rules are not obeyed by some persons and to some extent, the institution in question does not effectively exist and is not worth studying. Accordingly the question of conformity is fundamental to much of social science.

How to explain why people—all people at some time

or other—obey rules? If one thing is clear it is that rules are not obeyed automatically: they are not obeyed just because they are rules. We know this from experience in our own societies, and Malinowski taught us long ago in one of the great books of social science that it is just as true of the so-called primitive societies. Speaking of the "compulsory obligations of one individual or group towards another individual or group" he wrote: "The fulfilment of such obligations is usually rewarded according to the measure of its perfection, while non-compliance is visited upon the remiss agent." [9] Should the rewards or punishments fail, the primitive man does not continue to comply: "Whenever the native can evade his obligations without the loss of prestige, or without the prospective loss of gain, he does so, exactly as a civilized business man would do." [10]

As two more recent writers on the subject of conformity put it: "The greater and more valued the reward, the oftener it is achieved through conformity behavior, the more conformist the behavior is likely to become, and the more likely it is to become a generalized way of behaving in new situations." [11] In other words, if conforming to a norm brings a man success in attaining a reward, he is likely to conform.

---

[9] B. Malinowski, *Crime and Custom in Savage Society* (Paterson, N.J.: Littlefield, Adams, 1959), p. 12. The book was originally published in 1926.

[10] *Ibid.*, p. 30.

[11] E. L. Walker and R. W. Heyns, *An Anatomy for Conformity* (Englewood Cliffs, N.J.: Prentice-Hall, 1962), p. 98.

He is the more likely to do so, the more valuable he finds the reward. His success and the value of his success on one occasion makes his conformity more probable on other, similar occasions. But all these statements about the particular action of conforming follow directly from the propositions about actions in general that we have called the propositions of behavioral psychology. That is, the behavior that makes possible the very existence of institutions, which are in turn the things of most interest to the more "social" of social scientists, can only be explained through psychological propositions.

We must remember that this general explanation simply accounts for what happens if a man does in fact find the results of conformity rewarding. Just why he finds them so is another question, and in particular cases it may be more difficult to answer. He may not himself, for instance, believe at all in the usefulness of the rule but obeys it nevertheless, because other people, his associates, who do believe in the rule, will reward him with approval for doing so.

Of course the rules of many institutions are enforced by other institutions, as the rules against counterfeiting and against conspiracies in restraint of trade are enforced in our society by the government. But the enforcement itself depends on the psychological propositions: the enforcing agency tries to arrange that obedience to the rules will be rewarding to people—for the avoidance of punishment is a reward. The agency cannot do its job unless, in turn, some of its members obey its own rules, and so on. Those great institutional structures that in modern societies seem so to dominate men as to be almost beyond their control rest ultimately on something we feel to be weak as water—individual human choices. True, it is a concatenation of choices, such that the choices of some men make it more likely that other men will choose to act in a particular way; but the foundation is still human choice.

Few norms maintain themselves forever. The letter of the law is always becoming a dead letter. But disobedience no more escapes the psychological propositions than obedience does. For the propositions also imply that if men find some action other than conformity to be more successful in getting them a reward, or in getting them a more valuable reward, than conformity does, they are apt to take the other action. Since the norms of society are always in potential conflict, the other action may be conformity to another, more highly valued norm. If many men disobey, they are on the way to killing the institution of which the norm is part. If many of them disobey in the same way —and their companionship in crime is apt to make their individual disobediences more probable—they may be on the way to founding a new institution, for what many men do, it is often said all ought to do. Or better, since men seldom change all the rules of an institution at once, they are changing an old institution rather than founding a new one, though in time they may change it beyond all recognition. The explanation of institutional change no more escapes psychology than institutional stability does. Indeed it escapes it less, for in social change no one can take conformity for granted.

To sum up this part of the argument: I have taken three problems of explanation in three different social sciences—the explanation of the action of a particular man in history, the explanation of the laws of supply and demand, together with one class of exception to the laws, in economics, and the explanation of conformity to institutional rules, a matter of particular importance to anthropology and sociology. I showed in each case that the explanations required the use of some of the same propositions in behavioral psychology, whether or not the fact was made explicit by the sciences in question. I agree that the demonstrations I have used are relatively simple ones; I do not have the space for anything more complicated. And I could easily have increased their number without carrying any more conviction. Few and simple as they are, they still have something to tell us. Since the appearance of some of the same propositions in a variety of different deductive systems is evidence of the generality of these propositions for the explanation of findings within a particular field, I infer that propositions of behavioral psychology are the general explanatory propositions in the field of human behavior, that is, the field of social science. This does not mean of course that they are the only propositions that will appear in the explanations. Particular deductive systems will require particular propositions about the effects of given conditions. Nor does it mean that they allow us to explain everything. We shall never be able to explain many things because we have, and can get, no adequate information about the given conditions within which the general propositions are to be applied. I argue only that when we think we can explain, our general principles turn out to be psychological.

OBJECTIONS TO PSYCHOLOGICAL
EXPLANATION

The position I have taken here does me no great credit for originality. I think it is not only true, but obviously true. Obvious or not, it is a position that many social scientists violently resist accepting, especially some sociologists and anthropologists who are concerned with studying societies as organized wholes at a particular period of time, and who therefore do not have to come to grips with social change. The psychological propositions are, after all, propositions about the behavior of individual men, and it is this about them that bothers the more "social" social scientists.

Let me take up some of the more obvious objections first. Behavioral psychology is not a psychology of materialism. Just because it is concerned with reward, there is no implication that the rewards of human action must be material. Persons concerned with spiritual values should be the last to deny that they are rewarding.

As I have already suggested, the claim that the propositions of behavioral psychology are the general explanatory propositions of social science does not in the least imply—and this would specially shock the anthropologists—that "human nature is the same the world over." What the claim does assume is this: that a relatively few general propositions hold good of human behavior, from which, under a great variety of given conditions, including the rules passed on to the younger generation of a particular society by their elders, a great variety of different forms of concrete behavior follows. For that matter, the way in which the younger generation learns the culture of the society, as distinguished from what it learns, is again explainable by the psychological propositions. But to say that a wide variety of particular findings can follow from a few general propositions under a similar variety of given conditions is merely to say what is true of any theory that is a real theory.

The claim made here does not imply a "great man" theory of social science, though some great men have made a big difference in human development—a fact that sociologists, many of whom have given little thought to history, are not always ready to admit. Nor does it assume that to explain social phenomena one must be prepared to account for the behavior of every single individual concerned. That would certainly be impossible, and in any case the psychological propositions themselves imply that large numbers of individuals of similar backgrounds and placed in similar circumstances—say, members of a middle class in a period of economic expansion—are apt to behave in similar ways. Accordingly it is often both convenient and justifiable to talk about the behavior of aggregates, like a middle class, and not about the individuals that make it up, especially when what is to be explained is a statistical finding, a central tendency for the aggregate. But the fact that many social scientists necessarily treat many of the phenomena they study as collective in this sense does not make the propositions they use in explanation any less psychological.[12]

Nor, more generally, does the fact that the psychological propositions refer to the behavior of individuals mean that they necessarily explain social phenomena in terms of personalities. A personality is that precariously integrated set of interrelated responses that makes a particular person different as a whole from others. But to explain many social phenomena we do not need to consider men as "whole men," as personalities in this sense. Suppose there is an increase in the amount of money in the hands of the general population without any corresponding increase in the supply of goods. To understand the effects of this condition we do not need to consider how men differ among themselves in their behavior, for instance, as husbands or fathers. No doubt they do differ, and their differences may make a differ-

---

[12] See G. C. Homans, "Contemporary Theory in Sociology," in R. E. L. Faris, ed., *Handbook of Modern Sociology* (Chicago: Rand-McNally, 1964), p. 970. I have used passages from this paper, with minor modifications, elsewhere in this chapter.

ence to their individual economic responses. But the differences so far cancel themselves out as to allow the general trend towards a rise in prices to stand out clear. Though we certainly need psychological propositions to explain the rise in prices, we certainly do not need to take personalities into account. For this particular explanatory purpose we can deal with "economic men," so to speak, and not with "whole men."

Indeed we need the psychological propositions to explain personality itself, if that, rather than social trends, is what interests us. It is the long-continued sequence of rewards and punishments following on particular actions in particular circumstances, a sequence building on an individual's special genetic inheritance and beginning with his earliest childhood, that shapes the unique thing we call his personality, the powerful and related tendencies we observe in him today. Since behavioral psychology deals with the effects of rewards and punishment, it accounts for the steps in the sequence, so far as we have enough information about a person's past, which he and we usually do not, to apply the psychology. Certainly there is nothing, for instance, in Freud—not his emphasis on the importance of experiences in childhood, nor his emphasis on unconscious behavior—that runs counter to the findings of behavioral psychology. Indeed a good case could be made that Freud was the first great behavioral psychologist, though some modern ones would reject the theoretical apparatus in which he stated his conclusions. At any rate, a personality theory is not the alternative to a "social" theory of human behavior. Both social trends and individual personality are to be explained by the same body of general propositions.

The position taken here makes no assumption that men are isolated individuals. It is wholly compatible with the doctrine that human behavior is now and always has been social as long as it has been human. What the position does assume is that the general propositions of psychology, which are propositions about the effects on human behavior of the results thereof, do not change when the results come from other men rather than from the physical environment. The position does not deny in the least that new phenomena emerge when two men interact, phenomena that are not present when a single man interacts with the physical environment. New phenomena of this sort are emerging all the time. All it claims is that the explanation of the new phenomena does not require any new general propositions, that is, any propositions other than psychological ones. It does not require any general propositions that are specifically "sociological," referring to the characteristics of groups or societies as such. I have given above several such explanations of emergent phenomena, and I have written a whole book to show that the simpler features of social interaction in groups can be explained psychologically.[18]

The classical opponent of the position taken here is the great French sociologist, Durkheim. Speaking of what he called "social facts," he wrote: "Since their

[18] G. C. Homans, *Social Behavior.*

essential characteristic consists in the power they possess of exercising, from outside, a pressure on individual consciousnesses, it must be that they do not derive from individual consciousnesses and that accordingly sociology is not a corollary of psychology." [14] We must remember that Durkheim was thinking of a somewhat different psychology from ours: today we should lay more emphasis on individual actions than on individual consciousness. But, even with the change in emphasis, Durkheim is correct in his facts. Social facts, in his sense, certainly exist and exercise a constraint on the behavior of individuals. The great example of a social fact is a social norm, and the norms of the groups to which they belong certainly constrain towards conformity the behavior of many individuals. The question is not that of the existence of constraint but of its explanation. We have already seen that conformity itself requires psychological explanation. The norm does not constrain automatically: individuals conform, when they do so, because they perceive it is to their net advantage to conform, and it is psychology that deals with the effect on behavior of perceived advantage. From his own social facts the very opposite conclusion to Durkheim's may be drawn: that sociology *is* a corollary of psychology at least in the sense that social phenomena require general psychological propositions for their explanation.

[14] E. Durkheim, *Les Règles de la méthode sociologique* (Paris: Alcan, 1927), pp. 124-25.

METHODOLOGICAL INDIVIDUALISM

The position taken here coincides with what is called
by philosophers of history "methodological individualism."

> According to this principle, the ultimate constitu-
> ents of the social world are individual people who act
> more or less appropriately in the light of their dispo-
> sitions and understanding of their situation. Every
> complex social situation, institution or event is the
> result of a particular configuration of individuals,
> their dispositions, situations, beliefs, and physical re-
> sources and environment. There may be unfinished or
> half-way explanations of large-scale social phenomena
> (say, inflation) in terms of other large-scale phenom-
> ena (say, full employment); but we shall not have
> arrived at rock-bottom explanations of large-scale
> phenomena until we have deduced an account of
> them from statements about the dispositions, beliefs,
> resources, and inter-relations of individuals.[15]

I do not believe that the issue can be resolved by
arguments about what is really ultimate, what is really
real. I, for one, am not going to back into the position
of denying the reality of social institutions. For many

[15] J. W. N. Watkins, "Historical Explanation in the Social
Sciences," in P. Gardiner, ed., *Theories of History*, p. 505.

purposes we often and usefully treat social organizations, such as manufacturing firms, as social actors in their own right, even when we know their acts to be resultants of complex chains of individual decisions. The question is not whether the individual is the ultimate reality or whether social behavior involves something more than the behavior of individuals. The question is, always, how social phenomena are to be explained.

It might be argued that, if the ultimate units of social behavior are men and their actions, then the general propositions used to explain social behavior must be propositions about men and their actions; that is, they must be what I have called psychological propositions. In short, "methodological individualism" entails "psychologism." But this argument fails to persuade some men that one would have expected to be ripe for persuasion. Karl Popper, for one, is a methodological individualist who denies that he is also, in this sense, a psychologist.[16]

Unless they are ready to argue that social phenomena cannot be explained at all by deductive systems employing general propositions, the opponents of methodological individualism are bound to put forward an al-

---

[16] K. R. Popper, *The Open Society and Its Enemies* (New York: Harper & Row, 1963), II, 89-99. In taking this position, I think Popper fails to understand the nature of psychology and, perhaps, of explanation itself. His view that psychological explanation cannot account for "the unintended social consequences of our actions" is, I believe, quite unwarranted.

ternative type of explanation, which they might call "methodological socialism." [17] I doubt that a decision between the two could ever be reached by abstract argument. It is conceivable that at some time in the future—perhaps tomorrow morning—a sociological proposition will be discovered that is general, insofar as it applies to all social groups or aggregates, that has great power in explaining social phenomena, and that cannot itself be derived from psychological propositions. If it were discovered, all argument would fall down before the fact. I am certainly not against sociologists' trying to discover such a proposition, and I can find no line of reasoning that will demonstrate, before the fact, that it will not be discovered. I just do not believe, extrapolating from past experience, that this is going to happen—but the future is a long time.

What we can speak of is the present. No one now contends that psychological propositions can be used to explain all social phenomena. Our usual ignorance of the given conditions in which the propositions are to be applied is sufficient in itself to prevent that. But they certainly can be used, as we have seen, to explain some social phenomena. The weakness, here and now, of those opposing the position taken here is that, while arguing in general terms that psychology will never explain social phenomena, they do not produce alternative kinds of explanation. Let them take the problems of explanation seriously. Specifically, let them stop

[17] See A. C. Danto, *Analytical Philosophy of History* (Cambridge: Cambridge University Press, 1965), pp. 257-84.

citing in their favor the facts of social emergence, social constraint, even of a social whole's being greater than the sum of its parts—indeed to this last phrase it is difficult to assign any definite meaning. Whatever facts these words refer to are conceded in advance. Let them rather take specific phenomena that, in their view, exemplify emergence, constraint, wholeness, or indeed anything else, and show how they would explain them, sketching out their deductive systems in some detail. Then we might be able to tell whether or not their logic was apt to be adequate, and whether they had managed to avoid using psychological propositions. Above all, we might be able to tell whether they had an alternative type of explanation, one using purely sociological general propositions. The fact is that the people who attack the position taken here never produce their alternative general propositions. They are always telling us that our type of attack will never work, while never producing one that will.

FUNCTIONAL EXPLANATION

But this last remark is not quite fair. Some anthropologists and sociologists believe that they can offer an alternative type of explanation, one that starts with an alternative type of general proposition. The proposition refers not to men but to societies and to the needs of these societies. It takes the following general form: For all societies, if a society is to survive (or remain in equilibrium), it must possess characteristics of type *x*. The various types are sometimes specified; they are called the functional prerequisites of a society, and this sort of explanation of social phenomena is called "functional" or "structural-functional" theory.[18]

I have written so much about the inadequacies of functional theory that I shall repeat myself too much if I take up the whole problem again.[19] Let me rather take up a single example, actually, as we shall see, an example that mixes functional with psychological explanation. I cited earlier as one kind of general proposition in sociology the statement that all societies are stratified. I said it possessed little explanatory power but rather cried out for explanation. How might a functionalist explain it? I trust I do not reproduce unfairly the argument put forward in a famous article by Kingsley Davis and Wilbert Moore.[20]

In order to survive or remain in equilibrium, a society must motivate its members to carry out the activities necessary to its survival. The more important the

---

[18] For critiques of functionalism, see C. G. Hempel, *Aspects of Scientific Explanation*, pp. 297-330; E. Nagel, *The Structure of Science* (New York: Harcourt, Brace & World, 1961), pp. 520-35; R. S. Rudner, *Philosophy of Social Science* (Englewood Cliffs, N.J.: Prentice-Hall, pp. 84-111.

[19] G. C. Homans, *Sentiments and Activities* (New York: The Free Press, 1962), pp. 22-35; and "Contemporary Theory in Sociology" in R. E. L. Faris, ed., *Handbook of Modern Sociology*, pp. 962-67.

[20] K. Davis and W. E. Moore, "Some Principles of Stratification," *American Sociological Review*, X (1945), 242-49.

activities—that is, the more crucial they are for the survival of the society—the greater the society's need to motivate members to carry them out. But the supply of persons able and willing to carry out these activities—that is, to fill the more important positions in society—is short. To ensure an adequate supply, the society must make the rewards for filling the more important positions greater than those for filling the less important ones. A stratification system does just this. Accordingly, if the society is a surviving society, or a society in equilibrium, it will have a stratification system. Q.E.D. As Davis and Moore put it: "Social inequality is thus an unconsciously evolved device by which societies insure that the most important positions are conscientiously filled by the most qualified persons." [21]

This last statement is a good example of the lengths to which functionalism will lead otherwise intelligent men. As Dante would have said: "Let's not talk about it, but look and pass by."

Instead, let us recognize first that this functional explanation does in fact contain a psychological proposition—that is why I call it a mixed explanation—though as usual the proposition was left unstated both in the original argument and in my reproduction of it. It is the assumption that an increase in the value of the reward of an activity increases the probability that a person will perform the activity. Second, it is difficult to show that certain activities are more crucial for the

[21] *Ibid.*, p. 248.

survival of a society than others. It is hard to show, for instance, for a medieval society, that the peasants were less crucial to its survival than the knights. Yet the peasants were much less well rewarded than the knights. Perhaps all else we need for criticism is a simple appeal to realism, another word for which is the truth. It is not in fact "society" that rewards its members but other members or organizations, and it is they and not society that have specifiable needs—or, better, things they want, even if they do not need them. And the more "important" members are not just in a position of receiving greater rewards from society than other members. They are often in a position to take them—and this is one reason why they are important. To take a brutal but far from unique example, English society in 1066 did not offer the Normans the position of forming its upper class in order that the society should survive or remain in equilibrium; the Normans took the position by force of arms.

A real explanation of stratification should at least make a beginning at accounting for the actual position of different groups in actual societies, and so let us ask ourselves, for instance, why physicians stand pretty high, on the average, in the American stratification system. It is difficult indeed to show that a society would not survive or remain in equilibrium without doctors. Many societies have, in effect, done just that. Something else must be at work. In the United States the demand for doctors' services is high, and it is a demand that comes from people, not society. People have always set a high value on the cure of their ills, and in the last century doctors have become increasingly able actually to effect cures. On the other hand, the supply of doctors, for reasons we need not go into, is relatively short. The functional explanation is quite right in emphasizing scarcity, and wrong only in implying that "society" feels the pinch, for surely the functionalists are not using the word "society" simply as shorthand for its members. If few can supply what many want, and want badly, the few can often get, by deliberate action or otherwise, high rewards from the many. The few are, in fact, in a position of superior power. And if they command relatively high rewards and high power, other members of society will sooner or later perceive them as ranking high socially. The same is true of persons and groups controlling other forms of scarce services—that is, scarce and in great demand in the particular society in question—such as spiritual rewards or physical force (his control of force enables a man to do others a great service, for the power to kill is also the power to spare). As Gerhard Lenski rightly puts it: "The distribution of rewards in a society is a function of the distribution of power, not of system needs." [22] The reason why all societies are stratified is that in all societies persons and groups differ in power, and in turn the reason for this is that, finally, persons themselves are different, by birth or breeding, in their abilities.

[22] G. Lenski, *Power and Privilege* (New York: McGraw-Hill, 1966), p. 63.

I do not claim that this sketch takes into account everything necessary to explain stratification, but I am sure that it points in the main and right direction. In it, the unrealistic functional reference to society is replaced by a reference to men. And the general propositions when made explicit become propositions about the behavior of men. For the explanation of the effects of power, of the effects of high demand for services in short supply, certainly requires psychological propositions.[23] In short, the type of proposition that made the original explanation specifically functional simply disappears: it is unnecessary. I am sure that this could be shown to hold good of other functional explanations— if they were available for criticism.

Just as philosophy can give no reasons *a priori* for rejecting methodological socialism, so it can give no reasons for rejecting functional explanations. In principle, there could indeed be functional theories that met all the requirements of genuine explanation.[24] But are there in fact such theories in anthropology and sociology? Again, let the functionalists take the problem of explanation seriously and begin to spell out their deductive systems. Then we might be able to tell whether the systems were likely to explain something. The fact

[23] See, for examples, J. W. Thibaut and H. H. Kelley, *The Social Psychology of Groups* (New York: Wiley, 1959), pp. 100-25; G. C. Homans, *Social Behavior*, pp. 65-67; G. C. Homans, "A Theory of Social Interaction," *Transactions of the Fifth World Congress of Sociology* (International Sociological Association, 1964), IV, 113-31.

[24] See R. B. Braithwaite, *Scientific Explanation* (Cambridge: Cambridge University Press, 1953), Chapter 10.

is that the functionalists seldom do so, but set up schemes of classification and stop there. Davis and Moore come closer to spelling out their explanation than do the others, and we have seen what difficulties they get into.

THE EFFECTS OF FAMILIARITY

I turn now to a combination of other reasons, less often stated but none the less powerful for that, for the reluctance to accept the general propositions of social science for what they are. The first thing that gets in the way is our very notion of what science *is*, derived from the history of sciences like physics that became successful early. A scientist is someone who makes discoveries, who finds out something new. It is true that in one sense the early physicists did not discover anything that people did not know already. Long before Galileo made his experiment, they knew that balls rolled down slopes, and that apples fell off trees, long before one of them hit Newton. But even here the physicists made discoveries. Before Galileo and Newton, no one knew that the distance the balls rolled was a function of the square of the time or that the force on the apple could be measured by the product of two masses divided by the square of the distance between them. (Note that this kind of discovery, the discovery of precise mathematical relationships, is just the kind that social science has most trouble in making.) Certainly in the further reaches of physics and biology, the business of science has increasingly been to make discoveries that ordinary human experience had never come near to making. And in many areas what was discovered were precisely the more fundamental things, like the structure of the atom. As it is actual experience that makes expectations, the assumption grew up that this is what a science always does—that if a science does not make discoveries, and fundamental ones at that, it is hardly a science at all.

So far I have not stressed the most obvious difference between the social sciences and the other sciences. It is, of course, that they deal with the behavior of men. In no other science do the scientists study the behavior of things like themselves. The chief consequence of this condition, we often hear, is that social science can never become scientific, because the scientists, being men and thus holding strong feelings about human behavior, will let their emotions get in the way of their objectivity. I do not believe the danger to be as great as asserted. No doubt their emotions will lead them to study certain phenomena and not others. There are social scientists who prefer to study the things in society they like, and others who are positively compelled to study things they don't. But no great harm is done: the evidence is that either group gets great pleasure from pointing out to the other what it has deliberately left out, so that no trifle remains forever unconsidered. Nor is there any doubt that different social scientists will evaluate differently the conditions they do investigate, but with the present degree of acceptance of scientific standards, they will have less difficulty in agreeing on what the conditions *are*. We can all agree that American military officers on retirement often take important jobs in private industry, without all agreeing that this kind of link between government and industry constitutes—and here comes the evaluation—a conspiracy against democracy.

Much less often mentioned and at least as important is another kind of consequence of the unique position of social science. In studying human behavior social scientists are studying phenomena that human beings are uniquely familiar with.[25] As social animals, nothing has been more important to them, and there is nothing that they have learned more about. Since they are also talking animals and can pass on their knowledge, the knowledge has been accumulating for hundreds of thousands of years. They even know the fundamentals in the form of the principles of behavioral psychology. Though we surely cannot rule out the possibility that principles still more fundamental will be discovered in the future, these are fundamental enough. The ordinary man knows them in the sense that he is hardly astounded—on the contrary—when they are stated to him, even though he himself would not state them in the language of behavioral psychology, and though he has been known to be astounded

[25] See A. C. Danto, *Analytical Philosophy of History*, pp. 242-43.

by some of their ulterior implications in psychopathology. Above all he acts on them, not always successfully, in planning his own behavior.

All this familiarity has bred contempt, a contempt that has got in the way of the development of social science. Its fundamental propositions seem so obvious as to be boring, and an intellectual, by definition a wit and a man of the world, will go to any mad lengths to avoid the obvious. Add to this the dilemma created by the assumption that making fundamental discoveries is the mark of a science. Then either, if its fundamental propositions are already well known and so need not be discovered, social science cannot be a science, or, if it is a science, its principles must remain to be discovered and so must be other than these. Both views make trouble. In order to avoid mentioning the principles, social science may abandon the standards of science in explanation, or it may look for its fundamental principles in the wrong places and hence without success. The most significant difference between social science and other science is that its principles do not have to be discovered but, what is much more difficult, simply recognized for what they are.

Yet I think that accepting the view that the fundamental propositions of all the social sciences are the same, and that the propositions are psychological, would be a great advantage to all of these sciences. True, some of them will have to give up their cherished intellectual independence. Sociology, as the latest of them to receive academic recognition and thus anxious to make up for lost status, has gone so far, from Durkheim onwards, as to claim, not only that it possesses a type of theory all its own, but also that from its theory the findings of all the other social sciences follow as "interesting special cases." Yet by now sociology has enough real achievements to its credit that it needs no longer this artificial fillip to morale. Instead of talking like a *parvenu* let it act like an aristocrat—by not worrying about its status. The position taken here does not entail sociologists' doing anything different from what they do now, except giving up a kind of theorizing that they would be well rid of in any case.

In return for their lost independence, the social sciences would have much to gain from accepting the view presented here. I believe it to be the truth, and the truth is no small thing to lay hold on. Nor, though it is discovered by human minds, is it the mere creation of human minds. It lies out there, independent of us, untamable by us, in the nature of the universe.

To speak more practically, the acceptance of the view might mean that the solutions each of the social sciences has reached in dealing with its particular problems could be seen as relevant to, and contributing to the solution of, the problems of the others. This mutual support has already proceeded far, as for instance in economic and social history, but it has proceeded piecemeal, neither as far nor as fast as it might. A recognition that they all share the same principles might speed up the process. In short, what the social sciences have to gain is nothing less than intellectual unity.[26] We might even teach our different subjects with less waste of the student's time, since we should not have to ask him to make a fresh start with each one.

[26] For an interesting and important treatment of the social sciences as a single science see A. Kuhn, *The Study of Society: A Unified Approach* (Homewood, Ill.: The Dorsey Press, 1963).

3

THE DIFFICULTIES
OF EXPLANATION

The difficulties of social science lie in explanation rather than discovery. Explanation is the process of showing how empirical findings follow from, can be deduced from, general propositions under particular given conditions. The general propositions of all the social sciences are psychological, propositions about the behavior of men rather than about societies or other social groups as such. Although a time may come when these propositions will be shown in turn to follow from still more general propositions—physiological ones, for instance—this certainly cannot be done now.

These propositions are not only very general but also, in the sense I have already described, very well known. This fact supports the contention that the emphasis of social science must be rather different from that of many other sciences. If the fundamental propositions are already known, social science will put less energy into discovery, at least of fundamentals, and more into showing how the myriads of empirical findings follow from the fundamentals. I may have given the impression that this is easy—but then I chose easy, though important, examples. In fact it may in general be more difficult for the social sciences than for the others. Even if the propositions of behavioral psychology were accepted as the most general in the social sciences—and their very obviousness gets in the way of their acceptance at present—the difficulties with explanation that these sciences encounter would still be far from over. There may be propositions so general that they are of no use. To this problem I turn in the last chapter.

REDUCTION

One thing ought to be clear from my discussion of psychological propositions and other propositions. In its actual findings, social science operates at two levels at least: it possesses propositions about individuals and also propositions about aggregates—classes, groups, organizations, societies—made up of individuals. Besides the proposition that all societies are stratified, one example of the latter class might be the proposition that, in all highly industrialized societies, the characteristic family organization is the small or nuclear family rather than the large, extended, or joint family. Now in the broadest possible analogy, the same is true of many other sciences. Thus physics and chemistry possess propositions about atoms—we need not get down to the subatomic particles—and also propositions about aggregates of atoms: the masses in motion of mechanics.

My first point is that the relation between the two kinds of proposition is much more comfortable for physics than it is for social science. In mechanics we can treat the mass of a solid as concentrated at a single point and disregard the behavior of the individual atoms. In social science we can often do something a little similar: we can treat an organization as if it were an individual actor, but we recognize that this is at best a first approximation to reality and that if we wanted to explain just why an organization acted as it did, why it acted differently from other apparently similar organizations, we could not afford to disregard individuals. Nor could we explain why an organization like an army was ever defeated, short of physical annihilation. But armies are so defeated, and the reason is that individuals start to run away. The defeat of a group is its dissolution into its individual components.

And even generals, who are individuals too, lose the will to fight on.

The nature of the problem is perhaps clearer in thermodynamics. From its earliest achievement in Boyle's Law, thermodynamics was successful in stating and testing propositions about aggregates. Its basic variables—pressure, volume, and temperature—all referred to aggregates, such as gas in an enclosed space, rather than to the individual atoms and molecules making up the gas. And the propositions relating these variables to one another possessed both great generality and great explanatory power. Under a wide variety of given conditions, a wide variety of testable empirical conclusions could be drawn from them.

Later Willard Gibbs in his *Statistical Mechanics* showed that these aggregative propositions could in turn be derived from propositions about individual molecules, on the assumption that the latter behaved mechanically, a little like hard, round balls each possessed of mass, velocity, and direction and bouncing against one another. (I am, of course, greatly vulgarizing Gibbs's achievement.) It was unnecessary, and of course impossible, to follow the path of any individual molecule because, under conditions that were often realized naturally and could be standardized experimentally, the total behavior of individuals would tend to converge towards the measures of pressure and temperature for the aggregate.[1]

[1] For fuller discussion of reduction in thermodynamics see E. Nagel, *The Structure of Science* (New York: Harcourt, Brace & World, 1961), pp. 338-45.

What Gibbs did is often called *reduction,* but reduction is not different from deduction, except in the special sense of the deduction of the propositions of one named science from those of another. Gibbs reduced thermodynamics to statistical mechanics by showing that the aggregative propositions of the former could, under specific assumptions, be deduced from, derived from, the individualistic propositions of the latter—propositions, that is, about individuals. But note that Gibbs could not have become a reductionist until the development of thermodynamics had presented him with propositions, and pretty general propositions, to reduce. Let me put the matter in another way: the problem of explanation in thermodynamics was simplified by the fact that many of the empirical findings could first be derived from the very general aggregative propositions, and then the latter derived from the individualistic propositions. One did not have to make the direct jump from the most general to the most empirical.

The question is whether an analogous situation exists in social science. I have been accused of being a psychological reductionist in the sense of reducing the propositions of one named science, sociology, to those of another named science, psychology, but I doubt that I am a reductionist after the fashion of Gibbs. To be his kind of reductionist in the social sciences one must believe that there are propositions commanding both great generality and great explanatory power that hold good of social aggregates. I am not sure I do believe that this is true of many of the fields of social science. There certainly are propositions that can be explained, but where is their generality? Or if they are general, like the proposition about stratification, where is their explanatory power?

The analogy may come closest to holding good of economics. Some of the variables of economics, like prices, the volume of money in circulation, and the supply of goods, resemble pressure, temperature, and volume in thermodynamics in that they refer, not to the behavior of individuals, but to the aggregative resultants of the behavior of many individuals. The propositions relating these variables to one another have some generality and much explanatory power. And they certainly can be reduced to psychological propositions.

But, as has long been recognized, the very limitations of elementary economics gave it special advantages, allowing it to develop great power within these limits. It could work with continuous and easily measured variables: money prices and the quantities of commodities bought and sold. It could assume the existence of perfect markets, in which the behavior of no single trader could have much effect on the total result in prices or volume of trade, and in which the relations between traders were relatively impersonal, uncomplicated, that is, by relations between the individuals other than economic ones. Above all, elementary economics could take for granted, could simply take as given, the institutional arrangements and the larger social organization that maintained and made possible the other conditions.

The other social sciences possess few of these advantages, and in them it is hard to discover the sort of proposition that would allow the analogy with reduction in thermodynamics to be maintained. As I have often said, their tested propositions are many but only statistically true and usually very limited in scope. Our frequent experience in social science is to find that a proposition holding good in one set of circumstances does not hold good in another. It is not that the second set of circumstances masks the truth of the proposition. This situation occurs in physics, as when a feather floats in the breeze in apparent defiance of the laws of gravitation. It is rather that a different proposition, sometimes even the reverse of the first, holds good.

It would be risky to say that this situation will continue forever in all the social sciences. It is conceivable, for instance, that we shall develop a set of propositions about formal organizations, bureaucracies, that will be aggregative, in the sense of making no direct reference to individuals, but that will still possess wide explanatory power. An example of such a proposition might be: The more complex the division of labor in an organization, the larger the number of steps in its chain of command. But the process has certainly not gone very far. I suspect that history will be the last to succumb, particularly in dealing with problems it thinks of as peculiarly its own, like political history. Most historians are satisfied that they have discovered no very general propositions about historical processes as distinguished from propositions about the human actors in the processes. That, I think, is what they mean when they say that history does not repeat itself. If they ever do discover such propositions, they are apt to be of the kind that possess little explanatory power and themselves cry out for explanation. The issue for the social sciences is not whether we should be reductionists, but rather, if we were reductionists, whether we could find any propositions to reduce.[2] For most of these sciences and for the immediate future, the gap between the general, psychological propositions and the many empirical findings is likely to remain. Under these circumstances, explanation is difficult because complicated.

Strangely enough, if most of the social sciences did possess widely general propositions about social aggregates, they would be less worried about the question of reduction. After all, thermodynamics can handle most of its problems of explanation and prediction without ever calling for the help of statistical mechanics. The fact that the former can be reduced to the latter is of greater intellectual than practical interest: perhaps Gibbs's performance was no more than a marvelous stunt. In the same way, though Newtonian mechanics can be reduced to Einsteinian relativity, we still use Newton and not Einstein in send-

[2] "But, of course, for us to speak of reductions in the scientific sense, it is first required that we have two theories" (A. C. Danto, *Analytical Philosophy of History* [Cambridge: Cambridge University Press, 1965], p. 280).

ing rockets to the moon: the calculations are simpler and the difference of accuracy on the moon is of the order of one yard. Something of the same sort has been true of elementary economics: it has done so well at its own level that it has not had to worry about its intellectual underpinnings. Scientists argue about reductionism as an abstract issue only when it is not clear what is to be reduced and how it is to be done.

THE ORGANIC ANALOGY

At this point some persons will ask: You have tried the analogy with physical science and it has failed, but why not try the one with biological science? Perhaps a society is more like a living body than a gas? This is, in effect, the position of the functionalists in anthropology and sociology.

Organic bodies are highly integrated, their parts linked intricately together; the forces making for equilibrium are accordingly strong; and when equilibrium breaks down it is apt to do so with a bang: life and death are easy to tell apart, and provide a clear criterion of the viability of the organism as a whole. Physiology can make very general statements about the aggregates making up organic systems. It is able to tell us, for instance, what the kidney contributes to the functioning of the body without worrying very much about the individual cells of which the kidney is composed. (Of course there are other physiologists that do concern themselves with the nature of the cells.) Unlike those of physics, and like those of social science, the propositions of physiology are not always quantitatively very precise, but they have the alternative advantage that many of them come close to being of the all-or-nothing type, with sharp breaks in the values of the variables. If one loses one's kidneys, it makes, so to speak, not just a quantitative but a qualitative difference. Propositions like this are relatively, though only relatively, easy to establish, and with their help it is often possible to explain some clear and final result like death by spelling out the chain of relationships that leads, for instance, from the stab in the heart to the end of breathing.

All our dearest dreams to the contrary, we are not, when we deal with a group or society, dealing with a system like this. The social system is not as tightly integrated: a change in the other features of a society does not necessarily, for instance, mean a change in the family organization. The equilibrium forces are not as strong, and it is usually not at all clear when equilibrium breaks down. Beyond certain narrow limits of internal change a mouse body will simply die: it certainly cannot change into a hippopotamus. It is surprisingly hard to specify what constitutes the death of a society: the Assyrian empire is gone, but there are people, I hear, that still call themselves Assyrians. The organic analogy fails all along the line.

Social science gets the worst of both worlds. Its subject matter is remarkable neither for the relative simplicity of pre-atomic physics nor for the organic integration of physiology; its propositions are remarkable neither for quantitative precision nor for marked discontinuities. It is not easy to work out what contribution some social part makes to the integrity of a social whole, since the whole is so fluid. For purposes of discovery and explanation, these are all in their different ways great disadvantages.

If some of the social sciences seem to have made little progress, at least in the direction of generalizing and explanatory science, the reason lies neither in lack of intelligence on the part of the scientists nor in the newness of the subject as an academic discipline. It lies rather in what is out there in the world of nature. The human mind is not altogether a free agent. It is subdued to the material on which it works. In some parts of nature the material is more intractable than it is in others. Indeed I sometimes think that the nature of the material has been the cause of what looks like a lack of intelligence. If some sociologists have said some silly things in the matter of theory, I can readily forgive them. Sensing how difficult it would be to deal with the subject matter in the obvious and straightforward way, they have looked for short cuts. There is nothing wrong with looking for short cuts: it is a mark of intelligence to try to find them. And it is a mark of courage, but not of intelligence, to persist in looking for them when they are not there.

HISTORICITY

The question now is why, in social science, there should be such a gap between our general propositions, which are psychological and refer to individuals, and our propositions about aggregates, which are either of limited scope or of low explanatory power. One reason is obvious. Implied in the psychological propositions themselves is a strong element of historicity: past history combined with present circumstances determines behavior. Many of our propositions are limited by the variety of special historical circumstances in which they hold good.

The historicity, moreover, is double, lying both in the individual and in the group. For the individual the psychological propositions imply, for instance, that his past history of success in his activities under given circumstances determines whether he will try them again, or others like them, in similar circumstances. The reason why people with similar backgrounds behave in similar ways here and now—so far as they do behave in similar ways—is that they are likely to have had similar past experiences. But it is also history that makes our findings even at the level closest to individual behavior only statistically true. Many subjects with roughly similar backgrounds—say, American college freshmen—may respond in much the same ways to the experimental manipulations of small-group research, but a man differing even slightly from the rest in his past history may respond differently and upset the perfection of our correlations.

This historicity also, of course, holds good of groups and societies, large and small. The older generation of any group teaches the younger how it ought to behave. Though the elders may have lost confidence in some of their doctrines, they cannot go so far as to leave the young alone. True, the argument that "the socialization process" automatically prepares the young to take just those roles society is ready to offer them is a facile solution to the problem of social continuity.[3] A society may teach its young men values subversive to its own stability. Let a boy learn at his mother's knee such impeccable social values as independence and achievement, and he may, when he grows up and if he finds the opportunity, take action that will change the society radically, not maintain it. Still, some initial tendency for the young to learn, and thus perpetuate, the customs of the past will undoubtedly be there. This tendency for the past history of societies to influence their present character is strengthened by the fact that, even in their most violent revolutions, the members of a society either cannot change, or have no interest in changing, all their institutions at once. Some they wittingly or unwittingly preserve, so that the way change takes place is modified by the adherence of individuals and groups to other rules, which have not

[3] See D. H. Wrong, "The Oversocialized Conception of Man in Modern Sociology," *American Sociological Review*, XXVI (1961), 183-93.

changed. The past affects the very way in which the future comes into being.

There are many sciences that, for purposes of discovery and explanation, are not much affected by historicity. In explaining the effect of a lever in mechanics, one does not need to take into account the past history of the material of which it is made. One need only know that it is strong enough. Nor to understand the functions of the kidney does it make much difference what may have been the past history of its cells. But it is well to remember that for some parts of some sciences historicity does make a difference, and then these vaunted exact sciences are in no better shape than the social sciences.

I remember well when I first realized this. During World War II I was commanding a mine sweeper equipped to blow up magnetic mines harmlessly by exposing them to a magnetic field created between two long electrodes towed astern. Under these circumstances it was vital that the magnetic field of the sweeper itself should be of negligible force. Otherwise the mine would explode, shall we say, prematurely. I discovered that the science of electromagnetism could not predict or, *a fortiori*, explain what the field under the ship was going to be. It was not that the scientists did not have full confidence that they knew the laws of electromagnetism but that they found it difficult to draw specific conclusions from these laws under conditions in which the past history of the ship made a difference. For magnetic purposes a ship may be considered a piece of soft iron, and a peculiar property of iron is that various conditions can create in it a magnetic field which it will later retain. A ship's present magnetic field depends on the latitude and longitude of the place where she was built and on her heading on the builder's ways. It depends on where she has sailed and the courses she has steered during her life as a ship. It depends on the amount of pounding she has received from workmen and the sea. The individual mechanisms appear to be well understood. The difficulty lies in getting the information that will allow their application, and, if the information is available, of calculating the resultant of their interactions over time. The scientists could not have solved the problem even if they had wanted to, which they did not, for their practical purposes could be served, not by predicting the magnetic field of the ship, but by simply measuring it and neutralizing it, through the procedure called "degaussing."

I take this to be a parable of our problem in much of social science. Even if we had full confidence in our general propositions, the equivalent of the laws of electromagnetism, we should still have trouble in showing how our empirical propositions, the equivalent of the magnetic field of the ship, followed from them. Explaining, for instance, why the jury in its modern usage is one of our legal institutions is of the same order of difficulty. And the reasons for the difficulty are the same. It is often said that social science has been slow to make progress because the variables entering its problems are many and not easily controlled. But the basic

variables, those entering the general propositions of behavioral psychology, may be few. The difficulty does not
lie in the number of variables, but in the number of
men and groups in whose different activities the variables take different values. It lies above all in showing
how the behavior of different men, behavior exempli-
fying the same general propositions, combines over
time to produce particular results, when past behaviors
affect present ones in complex chains. It lies in the
historical or, as some philosophers of science say, the
genetic character of many of our explanations.

Let me, at the risk of repetition, try to put the matter in another way. Even in physical science, many explanations could be shown to involve history. Go back
to the lever I spoke of earlier. Explanations of the action of the lever tacitly take as given that the lever is
rigid and strong enough to bear without breaking the
forces put upon it. But if one were to try to explain in
turn why this given condition held good of a particular
lever, one would have to consider its history. What
material was it made of? If it is wooden, have age and
damp made it rotten? And so on. We do not usually
bother to ask this kind of question because the particular historical paths that lead to levers' being strong
enough are many and varied. And whatever the paths
may be, all we need to know is a general end-result.
That given, the action of any lever follows directly
from general principles. But this is not true of many of
the problems of most interest to social science. To explain an empirical finding, we feel the need to know

how a very specific given condition came about, which could not have been reached by any number of different paths but only by one or two very specific ones; for if this given condition had been different, the finding would have been different.

I think the question is finally one of intellectual satisfaction. To return to Scriven's example in the last chapter, it is easy to explain why William the Conqueror did not try to conquer Scotland, once the fact that William did not find its conquest valuable is taken as given. Though the explanation can easily be shown to require a general proposition, it is still uninteresting. Our interest only begins to be aroused when this very specific given condition is itself explained, when we have some idea how it came about. In general, the deductive systems of history are uninteresting when taken separately. Only when they are linked together in genetic chains, the "givens" of one deductive system becoming the *explicanda* of others, do they even begin to give intellectual satisfaction. But then the chains become complicated and difficult to establish.

There are, of course, physical and biological sciences, such as cosmology, historical geology, and paleontology, in which historicity makes as much trouble as it does in the social sciences. But for many of these sciences, including the most fundamental, it is not a problem. When it is one, the scientists may still be able to disregard it for practical purposes, since they are in a position to neutralize its effects. In most of the social sciences, on the other hand, historicity either in individuals or in groups is a problem right from the beginning, and the scientists are rarely able to neutralize it. I sometimes think that the social sciences are criticized as sciences for failing to do what a respectable physical science would not have even tried to do.

Why are the social scientists so stupid as to try? The reason lies again in the unique characteristic of social science: the scientists are studying the behavior of things like themselves. Nothing fascinates them more, and the more the others are like themselves, the more fascinated they become, because they hope to understand themselves and the groups they belong to. However intractable the problems, however small the theoretical or practical results to be achieved by solving them, there will always be an Englishman to explain the origins of Parliament, always a Southerner to explain why Lee lost the battle of Gettysburg.

DIVERGENT PHENOMENA

The historicity of social phenomena not only allows time, literally, for complexity to develop but also gives scope for divergence, the kind of thing we describe in the proverb, "For want of a nail the shoe was lost, for want of a shoe the horse was lost, for want of a horse the battle was lost," etc.[4] Or, "As the twig is bent, so

[4] See I. Langmuir, "Science, Common Sense, and Decency," *Science*, XCVII (1943), 1-7. Also W. Leontief, "When Should History Be Written Backwards?" *Economic History Review*, 2nd ser., XVI (1963), 1-8.

is the tree." In divergence, a force weak in itself but just tipping the scales in a balance of stronger forces has big and spreading effects over time. For instance, American institutions are in origin English institutions; English institutions would have been different if the Normans had not won England in 1066; and the Normans only just won the battle of Hastings after a whole day of very even fighting. It is divergence that makes much of the trouble in genetic explanations. It may be hard to get information about little things that made all the difference. And much explanation is convincing to the degree that weak, secondary forces can be neglected. In divergence they cannot be neglected.

Even when the same processes are occurring in different societies, divergence between them may be created by the different rates at which they occur. As we are beginning to learn, relatively slow industrialization in a country already heavily populated—and it may be the heavy population that slows up the industrialization—may have very different effects from rapid industrialization in a lightly populated country. Indeed it was perhaps their relatively low population plus a mechanical technology, including especially good ships and guns, already well developed by the end of the Middle Ages, and not any Protestant ethic, that gave the nations of northwestern Europe the lead in the Industrial Revolution.[5] The effect of an expanding trade

[5] See L. White, Jr., *Medieval Technology and Social Change* (Oxford: Oxford University Press, 1962), pp. 128-29; C. M. Cipolla, *Guns, Sails and Empires* (New York: Random House, 1965).

on limited manpower gave the West the incentive for applying labor-saving machinery to manufacturing. But why the West, compared with the East, had a relatively low population and an early lead in mechanics we may never know.

The social sciences share the same general propositions, but they differ in the degree to which they have to use the propositions to explain historical change of the divergent sort and cannot simply take its results as given. Explaining why particular institutions should be what they are and not something else is the most difficult and important problem, especially institutions in whose foundation and development particular individuals played a crucial part, as Henry II did in the development of the jury system or Hamilton in the establishment of the Constitution of the United States. If one can take institutions for granted, one is in pretty good shape. We have seen this to be true of elementary economics, but it is also true of much of sociology. To take a classic example, Durkheim's theory of suicide could explain why the suicide rate in Spain is low.[6] Suicide, he said, is positively related to individualism; Protestantism is more individualistic than Catholicism; and there are few Protestants in Spain. Durkheim—I am certainly not blaming him for doing so—simply took as given the facts that Protestantism, Catholicism, and Spain existed, that Protestantism was individualistic, and Spain, Catholic. But what is taken as given in

[6] E. Durkheim, *Suicide*, ed. G. Simpson (Glencoe, Ill.: The Free Press, 1947), pp. 152-70.

one deductive system may itself be explained by another. To explain the sorts of things Durkheim took as given is the really difficult problem. Though they do not require any special type of explanation, such as functionalism, the genetic chains that would account for them become terribly long and complex, and, unless cut off at some arbitrary point, lead back to the very beginnings of written history in the West. Though problems of this kind are perfectly legitimate, important, and interesting problems in social science, they are the most troublesome. The other social scientists leave them largely to the historians, while sneering at the unsystematic and untheoretical nature of history. The fact is that the materials historians have to work with, particularly in fields like political history that are most often abandoned to them, are just the kind of materials it is most difficult to get systematic and theoretical about.

CONVERGENT PHENOMENA

Fortunately, to ease the difficulties of explanation, convergence as well as divergence occurs in social phenomena, as when a number of groups or societies originally somewhat different wind up under the influence of strong forces at what is in some respects the same place. Such a convergent phenomenon is industrialization. The societies, whether the United States, Russia,

or Japan, that have successfully industrialized have undoubtedly converged in some ways, for instance, towards the nuclear organization of the family and towards a stratification system in which there are big
differences in status but no sharp lines to set off one
class as a whole from another. These societies have not
become by any means exactly the same; the remaining
differences between them are important, and they are
determined by the reverberating effects of the past his-
tories of these societies. Yet convergent phenomena ease
the difficulties of explanation by allowing us to neglect
secondary forces and small differences in timing.

It was, I think, because they wanted to escape the
real difficulties posed by historicity that the sociologists
and anthropologists attempted to develop a functional
theory: a theory that tried to explain the existence of
particular institutions by the contribution they made,
jointly or singly, to the survival or equilibrium of a society here and now. The anthropologists were the better
justified of the two, since, after all, most of the societies
they studied were nonliterate and possessed no written
records. Yet even anthropology cannot escape historical
explanations of the looser genetic form made possible
by convergent phenomena.

To convergent phenomena I am sure we owe some of
anthropology's most successful propositions about the
interrelations of institutions: the cross-cultural propositions about the interrelations of various features of
primitive kinship systems. These propositions are suc-

cessful in that, since the primitive societies are relatively numerous, they can be tested statistically, and they sum up much of the data. Though anthropologists are not agreed on how they are to be explained, I myself have no doubt that the explanations must be broadly genetic.

Take, for instance, the proposition that strongly patrilineal societies are apt to have institutionalized a close, warm, and free relationship between a man and his mother's brothers.[7] Once societies had adapted patrilinity for whatever reasons in the past, the forces thus set in motion must have been strong enough to make the societies converge towards the ego-mother's brother relationship, just as societies that industrialized set in motion forces tending towards a nuclear family system. Crudely and elliptically put, the argument is as follows: that in patrilineal societies fathers hold jural authority over their sons; that authority, because it restricts freedom and may be exercised through punishment, tends to inhibit a close, warm, and free relationship between the person in authority and the person submitted to it; that a man may nevertheless have strong needs that can be satisfied through such a relationship, especially with an older person of the same sex; and, finally, that the nearest such person in a patrilineal society, the nearest older man that does not have authority over him, is a man's mother's brother.

[7] See, for instance, G. C. Homans, *The Human Group* (New York: Harcourt, Brace & World, 1950), pp. 19c-261.

Note that the argument is inherently psychological. More important for my present purposes, it is broadly genetic. It is not historical, since it cannot appeal to historical records. Indeed it assumes that the historical details are not important. The process by which the institutionalized relationship between ego and mother's brother was arrived at may well have differed from society to society, but the forces making for convergence were strong enough to overcome the differences. Yet the argument is genetic in that it assumes that the development of patrilinity was a prerequisite of, a necessary stage towards, the development of the mother's brother–sister's son relationship.

But why should so many societies have chosen—if that is the right word—patrilinity in the first place? We may never know, but one possible guess is that keeping the men of a group together generation after generation in association with a particular territory may have offered real advantages in the economic conditions faced by many primitive tribes. One of the obvious ways of keeping them together is keeping fathers and sons together, and this is the essence of patrilinity. Even the anthropologists who would not agree that this is the correct explanation have no other to offer. I myself believe that many explanations in social science will turn out to be genetic in this sense. They will assume that one type of phenomenon must have been in process of occurring before convergent forces could have brought another type of phenomenon into being.

PSYCHOLOGY AND THE ILLUSION
OF CHOICE

In the social sciences our only general propositions are propositions about individual behavior. Many of our aggregative propositions are only statistically true and hold good only within particular historical circumstances. The reason is that the historicity of the phenomena, while allowing room for convergence, still allows plenty of room for divergence in the circumstances in question.

If these conditions make trouble for us as social scientists, remember that they are a great advantage to humanity, by leaving men the illusion of choice.[8] I speak of the illusion because I myself believe that what each of us does is absolutely determined. But it is one thing to believe in determinism as a matter of faith, and quite another thing to demonstrate it in detail. Think what it would take to construct a computer and program it, then to gather and feed into it the data that would enable it to predict ahead of time, not just some of the things a man would do, but just exactly what he would do in the course of a day. It would be even more difficult to predict if he knew ahead of time that his behavior was going to be predicted—yet a fair test of determinism would have to include this con-

[8] See A. C. Danto, *Analytical Philosophy of History*, pp. 283-84.

dition. Even if we could do this, the effort would cost more than the result would be worth, for I think it would tell us only what we know already. The illusion of free will is going to be saved by cost considerations, but it is a vital illusion. If there were—as there are not—a few macroscopic laws about society, rather than laws about individuals, that hold good across the board, mankind would lose the conviction, which some part of it, thank God, preserves, that it can by taking thought change its condition in ways it considers better—even if by its own standards not all the changes would turn out to be so. The most amusing case is that of the Marxists, who theoretically believe in macroscopic laws inevitably converging on a certain result, but who will not let the laws alone to produce the result, and insist on helping them along. Their humanity will keep breaking through, and in the crunch of reality they behave just like the rest of us, who believe, in the words of Mr. Justice Holmes, that "the way in which the inevitable comes to pass is through effort." [9]

EPILOGUE

Years ago Chester Barnard wrote: "It seems to me quite in order to cease encouraging the expectation that

[9] J. B. Peabody, ed., *The Holmes-Einstein Letters* (New York: St. Martin's Press, 1964), p. 5.

human behavior in society can be anything less than the most complex study to which our minds may be applied." [10] I hope I have done nothing to encourage the expectation. I have tried to point out the nature of the complexity and some of the reasons for it. If our results are limited, we should not excuse the fact with the argument that our science is young. It is not all that young, and it has been very energetic. Its difficulty is a better excuse than its youth, and one that does us more credit.

Indeed our results do not appear so limited if we look at the many empirical findings. Our trouble has not been with making discoveries but with organizing them theoretically—showing how they follow under a variety of given conditions from a few general principles. That is what I have meant by saying that the problem of social science is not discovery but explanation. Only if we recognize the nature of our problem can we begin to cope with it.

Don't tell me our problem lies in prediction. No doubt it will remain our greatest weakness. But prediction runs parallel to explanation: the two problems are really the same one. And the better we are able to explain what has happened, the better we shall be able to predict what will.

In explanation our difficulty does not lie in any lack of general propositions. On the contrary, we have known the general propositions, in one form or another,

[10] C. I. Barnard, *The Functions of the Executive* (Cambridge· Harvard University Press, 1938), p. xii.

so long and so well that we cannot believe they *are* general. Familiarity has bred contempt. If we are to make progress in the intellectual organization of our subject, our first job is to acknowledge the general propositions for what they are and to acknowledge that all the social sciences share the same ones. If we admit this kind of unity, we might be able to pool our resources with less jealousy.

Our difficulty in explanation lies in the nature of the general propositions. They are propositions about individual behavior, yet what the social sciences often want to explain are the characteristics of social aggregates. The central problem of social science remains the one posed, in his own language and in his own era, by Hobbes: How does the behavior of individuals create the characteristics of groups? That is, the central problem is not analysis but synthesis, not the discovery of fundamental principles, for they are already known, but the demonstration of how the general principles, exemplified in the behavior of many men and groups, combine over time to generate, maintain, and eventually change the more enduring social phenomena.[11]

In dealing with the problem of complex explanation, no good will be done by failing to recognize what ex-

[11] For an argument similar to the one advanced in this book, see F. Barth, "Models of Social Organization" (Royal Anthropological Institute of Great Britain and Ireland, Occasional Paper No. 23, 1966).

planation *is*. Only by asking how we actually explain social phenomena can we begin to understand what the
real difficulties of social science are. No good will be
done by treating theory as if it were something other
than explanation, by abandoning the standards of the
other sciences in explanation, as I think they are often
abandoned in theoretical work in sociology, or by writing one kind of "theory" for the record, while keeping
our actual explanations under the table. I cannot repeat too often that our actual explanations are our
actual theories. What is needed is the most utter intellectual honesty, including the humiliating honesty of
admitting the obvious.

If we recognize the nature of our intellectual problem, there is no other reason for despair. In many situations, and not only in economics, we can make much
progress in explaining the behavior of men by taking as
simply given, for the time being, the institutional structures within which they operate, even if their behavior
within the structures will eventually change the structures. In explaining institutions and the relations between them, our task is often made easier by the pres-
ence of powerful convergent processes. And in the most
difficult problem of synthesis, the problem of explaining the possibly divergent resultants of complex interactions between individuals and groups over time, the
high-speed computer has just arrived to help us. It
will not do everything for us; it won't ask the right
questions—that remains our job—but the answers, so

far as they depend on the mechanics of calculation, the computer will get for us with a speed no man can match.

If the central problem of the social sciences is to show how the behavior of individuals creates the characteristics of groups, we should pay particular attention to the situations in which social phenomena can be most convincingly explained by psychological propositions—convincingly, because in these situations we can directly observe the behavior of the individuals concerned and thus get the information necessary for explanation. This is the strategic justification of what is called small-group research. In the small group we can really observe and explain how conformity occurs, how power is exercised, and how status systems arise. These are surely among the most convergent of social phenomena, and ones we must begin by understanding intimately if we are ever to grasp the nature of larger societies. It is surprising how little, in any systematic way, we do understand them.

Though the general propositions of all the social sciences are propositions of behavioral psychology, we need not for that reason pay any special deference to the behavioral psychologists. They have been surprisingly unenterprising and even naïve in extending their propositions to the explanation of social behavior. Most of the work in the field of small groups has been done by social psychologists and sociologists who find —I think wrongly—that behavioral psychology is limited in its applicability: adequate for rats and pigeons but not for men. Let them acquire some of the tough-minded qualities of the behaviorists and employ them in a crucial field that the behaviorists have surprisingly left to them by default.

Much of social science resembles what social geographers call a "hollow frontier." In Brazil, for example, successive bands of adventurers, seeking gold, diamonds, or rubber, have swept through the backwoods, advancing the frontier but leaving behind it little in the way of settled territory, consolidated for civilization. So it is with social science. We adventurers reach more and more "exciting" findings; foundations give us more and more money to advance the "growing edge" of our fields, but behind the growing edge the intellectual work that would organize the findings remains largely undone. It can be done only if we take the job of explanation seriously—for to explain is to organize—and try to explain at least the more familiar features of social life.